Mathematical Foundations of Computer Science

Mathematical Foundations of Computer Science introduces students to the discrete mathematics needed later in their Computer Science coursework with theory of computation topics interleaved throughout. Students learn about mathematical concepts just in time to apply them to theory of computation ideas. For instance, sets motivate the study of finite automata, direct proof is practised using closure properties, induction is used to prove the language of an automaton, and contradiction is used to apply the pumping lemma.

The main content of the book starts with primitive data types such as sets and strings and ends with showing the undecidability of the halting problem. There are also appendix chapters on combinatorics, probability, elementary number theory, asymptotic notation, graphs, loop invariants, and recurrences. The content is laid out concisely with a heavy reliance on worked examples, of which there are over 250 in the book. Each chapter has exercises, totalling 550.

This class-tested textbook is targeted to intermediate Computer Science majors, and it is primarily intended for a discrete math / proofs course in a Computer Science major. It is also suitable for introductory theory of computation courses.

Ashwin Lall is Professor of Computer Science at Denison University. He joined the Denison faculty in 2010. Prior to this, he was a postdoctoral researcher at Georgia Tech, a Ph.D. student and Sproull fellow at the University of Rochester, and a math/computer science double major at Colgate University. Dr. Lall has taught all the existing flavors of the introductory Computer Science course as well as advanced topics such as Theory of Computation and Design/Analysis of Algorithms. He also enjoys teaching the Game Design elective in the CS major.

Mathematical Foundations of Computer Science

Ashwin Lall

CRC Press
Taylor & Francis Group
Boca Raton London New York

CRC Press is an imprint of the
Taylor & Francis Group, an **informa** business

A CHAPMAN & HALL BOOK

First edition published 2025
by CRC Press
2385 NW Executive Center Drive, Suite 320, Boca Raton FL 33431

and by CRC Press
4 Park Square, Milton Park, Abingdon, Oxon, OX14 4RN

CRC Press is an imprint of Taylor & Francis Group, LLC

© 2025 Ashwin Lall

ISBN: 978-1-032-46789-4 (hbk)
ISBN: 978-1-032-46787-0 (pbk)
ISBN: 978-1-003-38328-4 (ebk)

DOI: 10.1201/9781003383284

Typeset in Latin Modern font
by KnowledgeWorks Global Ltd.

Publisher's note: This book has been prepared from camera-ready copy provided by the authors.

Contents

Preface

The guiding principle behind this book is to make the mathematics needed in computer science accessible to students who are motivated by computer science topics. By being introduced to mathematical topics just in time to understand the next computer science concept, students will be motivated to master the mathematical ideas. This book also includes concepts such as automata, regular expressions, grammars, Turing machines, and computability, which students may not otherwise see until much later courses (if they see them at all).

I assume that the reader of this book has had an algebra course (not necessarily calculus) and has taken an introductory computer science course that covers computer arithmetic, functions, iteration, conditionals, and recursion.

NOTE FOR STUDENTS

This book is written to be example-heavy—there are over 250 worked examples. As you read each concept, work through the examples that come right after. (Keep paper and pen handy!) To solidify your understanding, practice the concept using the exercises that are given at the end of each chapter—there are over 550. The book was written with concision in mind, but this means that there are some sections you may have to read through several times to fully understand the ideas. It is well worth making sure you understand each section before moving on, as each section builds on the previous one(s).

NOTE FOR INSTRUCTORS

The goal of this book is to introduce students to the mathematical concepts needed in computer science just in time to apply them to learning some basic theory of computation. I have found that having real computer science topics helps motivate the mathematics for most students. The just-in-time philosophy works really well with theory of computation topics as you can make connections such as sets/languages to DFAs, direct proof to closure properties, induction to proving the language of a DFA, and proof by contradiction to the pumping lemma.

For programs that have a limited number of credits that they can require (such as at liberal arts colleges), this book offers a way to fold discrete math and automata theory into a single course. It covers most of the standard topics in both courses with some exceptions (e.g., pushdown automata). If you choose to adopt the approach of this book, your department will have to coordinate learning goals between this course and any later theory of computation course that you offer.

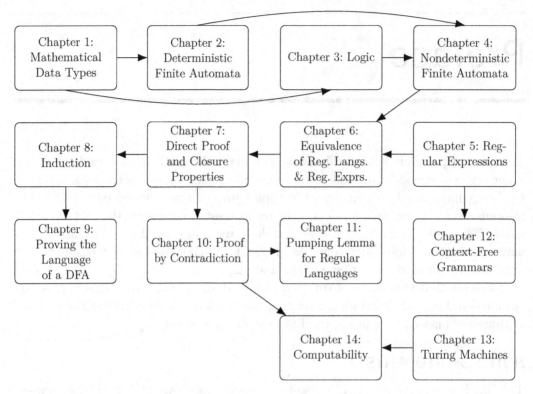

Figure 1 Chapter dependencies

While I recommend going through most chapters sequentially, please feel free to reorder some topics. The dependencies between chapters are given in Figure 1. I prefer to get to DFAs (Chapter 2) very quickly, but you could choose to cover logic (Chapter 3) before it. When I teach this course over 15 weeks, we typically get through the entire book with the exception of Chapters 9 and 12 and Appendices C and E. Depending on time constraints at your institution, you can also choose to not spend a lot of time on Chapter 6 and cut some of the later theory of computation material (Chapters 13 and 14), though it is compelling to get to decidability in this course. The appendices are included to cover other topics that are needed by your curriculum. They can be taught in any order, except that Appendix A (Counting) briefly references Chapter 3 (Logic), Appendix B (Probability) uses Appendix A, Appendix D (Asymptotic Notation) relies on Chapter 3 (Logic), and Appendices F and G (Loop Invariants and Recurrence Relations) should be covered after Chapter 8 (Induction).

This book was typeset in LaTeX, and all figures were made using TikZ. I recommend having your students write up their solutions in LaTeX. Many resources for this book can be found at https://personal.denison.edu/~lalla/MFCS.

If you do end up adopting this book for a course, I would love to hear about it at lalla@denison.edu. If you have comments or find any typos, please also feel free to reach out.

ACKNOWLEDGMENTS

I'd like to thank the following individuals:

- Randi Cohen and Solomon Pace-McCarrick, for guiding me through the process of getting this book published;

- Sean McCulloch and Sam Gutekunst, who gave me some excellent feedback that considerably improved the style of the book;

- Ali Miller, for using a draft of this book and giving me practical feedback for many parts;

- my students in the Fall 2023 section of CS 234, who gave me careful feedback: Caroline, Chloe, Daniel C., Lam, Khoa, Kian, Oghap, Ashwin, Phan Ahn, Phuong, Eleanor, Andrew, Cheryl, Tomer, Aryah, Ethan, Dipanker, Hieu, John, Daniel Z;

- my PhD advisor, Mitsu Ogihara, for his guidance on this project;

- the late Laura Sanchis, who instilled in me a love for the theory of computation that I hope is replicated in even a single reader of this book;

- my Denison colleagues for their support, both direct and indirect, particularly Dee Ghiloni, Mikey Goldweber, Dave Goodwin, Jessen Havill (now at Bucknell), David Kahn, Matt Kretchmar, Matt Law, Lew Ludwig, May Mei, Ali Miller, Tony Silveira, Stacey Truex, and David White;

- Pria and Ramona, for decades of love and care; and

- Erin, Kiran, and Shannon: the best family a dad-dad could ask for.

This book is dedicated to the memory of my father, Ajit.

CHAPTER 1

Mathematical Data Types

1.1 WHY YOU SHOULD CARE

In this chapter, you will learn about the basic mathematical objects that you will use throughout this book. You can think of this as analogous to learning about primitive data types (e.g., `int` or `float`) in a programming language. It is critical to understand these very well to be able to understand more complex ideas.

1.2 SETS

You can think of a set as a box with a collection of objects in it. For example, below is an instance of a set with the numbers 1, 2, and 3 in it. This set is represented by the notation $\{1, 2, 3\}$.

Example 1.1

The set consisting of 1, 2, and 3: [1 2 3]

There are a few important things to note about the idea of a set. First, a set does not have any repeated objects. If an object is repeated, it is only counted once. In particular, both the sets in the example below are the same because it doesn't matter how many times 3 appears in the set.

Example 1.2

The sets $\{1, 2, 3\}$ and $\{1, 2, 3, 3\}$ are the same: [1 2 3] [1 2 3 3]

Second, the order of a set is immaterial. The sets $\{1, 2, 3\}$ and $\{2, 1, 3\}$ are the same set. This is very different from the idea of a list in a programming language like Python, where $[1, 2, 3]$ and $[2, 1, 3]$ are different lists.

Sets can contain objects of a variety of types, not just numbers.

DOI: 10.1201/9781003383284-1

Example 1.3

The set $\{a, b, c, d\}$ contains the first four letters in the English alphabet. The set {Python, Java, C++} contains three programming languages.

We are now prepared to see our first definition:

Definition 1.1

A set is an unordered collection of distinct objects.

1.3 SET TERMINOLOGY

There is much terminology surrounding sets that is necessary to know as a computer scientist. The easiest way to learn it is to see lots of examples.

Any object in a set is called an **element** or **member** of the set.

Example 1.4

The number 2 is an element of $\{1, 2, 3\}$, written as $2 \in \{1, 2, 3\}$.

We say that any object not in a set is not an element.

Example 1.5

The number 4 is not an element of $\{1, 2, 3\}$, written as $4 \notin \{1, 2, 3\}$.

If all the elements of a set A are elements of another set B, we say that the first set is **contained** in the second or that A is a **subset** of B and write this as $A \subseteq B$.

Example 1.6

The set $\{1, 2, 3\}$ is contained in (is a subset of) the set $\{1, 2, 3, 4\}$, written as $\{1, 2, 3\} \subseteq \{1, 2, 3, 4\}$.

A special case of this is that a set is always a subset of itself.

Example 1.7

The set $\{1, 2, 3\}$ is a subset of $\{1, 2, 3\}$ ($\{1, 2, 3\} \subseteq \{1, 2, 3\}$).

A special set, called the **empty set**, denoted by $\emptyset$, is the set with no elements. The empty set is a subset of all sets, including itself.

Example 1.8

The set $\emptyset$ is a subset of $\{1, 2, 3\}$ ($\emptyset \subseteq \{1, 2, 3\}$).

A set is a **proper subset** of another if the former is a subset that does not have all the elements of the latter.

Example 1.9

The set $\{1, 2, 3\}$ is a proper subset of the set $\{1, 2, 3, 4\}$, written as $\{1, 2, 3\} \subset \{1, 2, 3, 4\}$.

Two sets are **equal** if they have exactly the same elements. An alternative definition is that both sets are subsets of each other.

Example 1.10

The set $\{1, 2, 3\}$ is equal to the set $\{3, 1, 2\}$.

The **cardinality** of a set is the number of distinct items it contains. The cardinality of a set is denoted by vertical bars around the set.

Example 1.11

The cardinalities of the sets $\{1, 2, 3\}$, $\emptyset$, $\{1, 2, 3, 2\}$ are $|\{1, 2, 3\}| = 3$, $|\emptyset| = 0$, and $|\{1, 2, 3, 2\}| = 3$.

A set is called **finite** if its size or cardinality is a finite integer.

Example 1.12

The sets $\{1, 2, 3\}$, $\emptyset$, $\{1, 2, \ldots, 100\}$ are all finite with sizes 3, 0, and 100, respectively.

A set is called **infinite** if the number of elements in it is infinite.

Example 1.13

The sets $\{1, 2, 3, \ldots\}$, and $\{\ldots, -2, -1, 0, 1, 2, \ldots\}$ are both infinite.

There are a few infinite sets that have special symbols to represent them. The set of natural numbers $\mathbb{N}$ is defined as:

Definition 1.2

$\mathbb{N} = \{0, 1, 2, 3, \ldots\}$.

The set of integers $\mathbb{Z}$ is defined as follows:

Definition 1.3

$\mathbb{Z} = \{\ldots, -3, -2, -1, 0, 1, 2, 3, \ldots\}$.

The set of all real numbers is denoted by $\mathbb{R}$.

Definition 1.4

$\mathbb{R}$ = all real numbers.

1.4 SET-BUILDER NOTATION

When we are representing sets, especially infinite ones, it is useful to use **set-builder notation**. In this notation, a set is given in the form {expression : constraints}, where the expression is an operation performed on all possible values with the given constraints before insertion into the set.

Example 1.14

The set $\{n : n$ is a natural number less than 5$\}$ is set-builder notation for $\{0, 1, 2, 3, 4\}$.

You can think of this as a `for` loop that iterates through all the values that fit the conditions to the right of the colon (i.e., the natural numbers 0, 1, 2, 3, 4) and inserts these values into the set.

The expression to the left of the colon is an operation that can be applied to the values being iterated over. In the above example, the operation does nothing to each value. In the example below we double the value instead.

Example 1.15

The set $\{2n : n$ is a natural number less than 5$\}$ is set-builder notation for the set $\{0, 2, 4, 6, 8\}$.

Set-builder can also be used for sets with infinite elements, even though a `for` loop to build it would never terminate.

Example 1.16

The set $\{n^2 : n \in \mathbb{N}\}$ is set-builder notation for the set $\{0, 1, 4, 9, 16, \ldots\}$.

Example 1.17

The set $\{2n : n \in \mathbb{Z}\}$ is set-builder notation for the set of even integers $\{\ldots, -4, -2, 0, 2, 4, \ldots\}$.

We can also use set-builder notation to describe sets using plain English, as shown in the examples below:

Example 1.18

The set of real numbers $\mathbb{R}$ can be written as $\{n : n$ is a real number$\}$.

Example 1.19

The set of rational numbers (numbers that can be written as a fraction) $\mathbb{Q}$ can be written as $\{p/q : p, q \in \mathbb{Z} \text{ and } q \neq 0\}$. Note that we disallow q to be zero to avoid division by zero and that there can be many ways to write a given fraction (e.g., $1/2 = 2/4 = 3/6 = \ldots$).

The type of elements of the set can sometimes be included before the colon, as shown in the example below:

Example 1.20

The set $\{x \in \mathbb{N} : 3 \leq x \leq 8\}$ is another way of writing the set $\{3, 4, 5, 6, 7, 8\}$.

1.5 UNION, INTERSECTION, DIFFERENCE, COMPLEMENT

There are a number of standard operations that are performed on sets. The **union** of sets is the set of elements in at least one of them.

Example 1.21

The union of $A = \{1, 2, 3\}$ and $B = \{2, 3, 4\}$ is written as $A \cup B = \{1, 2, 3, 4\}$.

Example 1.22

The union of $A = \{1, 2\}$, $B = \{2, 3\}$, and $C = \{3, 4\}$ is $A \cup B \cup C = \{1, 2, 3, 4\}$.

The **intersection** of sets is the set of elements in all of them.

Example 1.23

The intersection of $A = \{1, 2, 3\}$ and $B = \{2, 3, 4\}$ is written as $A \cap B$ and is equal to $\{2, 3\}$.

Example 1.24

The intersection of $A = \{1, 2, 3\}$, $B = \{2, 3, 4\}$, and $C = \{3, 4, 5\}$ is written as $A \cap B \cap C = \{3\}$.

The **difference** of two sets is the set of elements in the first that are not also in the second. The difference of sets A and B is written $A - B$ or $A \backslash B$.

Example 1.25

The difference of $A = \{1, 2, 3\}$ and $B = \{2, 3, 4\}$ is written as $A - B$ and is equal to $\{1\}$. Note that this is different from the difference $B - A = \{4\}$.

In some cases, sets are defined as being subsets of a universal set, sometimes denoted by the symbol U. When such a universal set is defined, we can define the **complement** of a set A to be the set of elements in U that are not in A (i.e., $U - A$).

Example 1.26

If we take the universal set to be $U = \{1, 2, 3, 4, 5, 6, 7, 8, 9, 10\}$, then the complement of $A = \{2, 3, 5, 7\}$ is written as $\overline{A}$ and is equal to $\{1, 4, 6, 8, 9, 10\}$.

Example 1.27

If we take the universal set to be $U = \mathbb{Z}$, then the complement of $A = \{x \in \mathbb{Z} : 1 \leq x \leq 100\}$ is $\overline{A} = \{x \in \mathbb{Z} : x < 1\} \cup \{x \in \mathbb{Z} : x > 100\}$.

Some writers use the notation A^c or $\sim A$ to denote complement, but we will use $\overline{A}$ throughout this book.

In terms of order of operations, the complement operation has the highest precedence and all other ordering is shown using parentheses.

Example 1.28

If $A = \{1, 2, 3\}, B = \{2, 4, 6\}, C = \{3, 4, 5\}$ and we assume a universe of $U = \{1, 2, 3, 4, 5, 6\}$, then $(\overline{A} \cup B) \cap C$ is $\{4, 5\}$ because $\overline{A} = \{4, 5, 6\}$ so $\overline{A} \cup B = \{2, 4, 5, 6\}$ and thus $(\overline{A} \cup B) \cap C = \{4, 5\}$. On the other hand, $\overline{A} \cup (B \cap C)$ is $\{4, 5, 6\}$ as $\overline{A} = \{4, 5, 6\}$ and $B \cap C = \{4\}$. Notice that it would not make sense to have an expression like $\overline{A} \cup B \cap C$ as it is ambiguous without the parentheses.

Example 1.29

If $A = \{1, 2, 3\}, B = \{2, 3, 4\}, C = \{3, 4, 5\}$ and we assume a universe of $U = \{1, 2, 3, 4, 5, 6\}$, then $(A - B) - C$ is $\{1\}$ as $A - B = \{1\}$. On the other hand, $A - (B - C) = \{1, 3\}$ as $B - C = \{2\}$. Note again that it would not make sense to have an expression like $A - B - C$ as it is ambiguous.

1.6 VENN DIAGRAMS

A way to visualize set operations is to draw Venn diagrams. These diagrams shade in the region representing the operation between one or more sets. Venn diagrams for the various set operations described in the previous section are given in Figure 1.1.

We can even visualize set operations involving three sets, as seen in Figure 1.2.

These Venn diagrams can sometimes be used to demonstrate relationships between different expressions. For example, De Morgan's Laws state that $\overline{A \cup B} = \overline{A} \cap \overline{B}$ and $\overline{A \cap B} = \overline{A} \cup \overline{B}$. These can be derived from Venn diagrams as seen next.

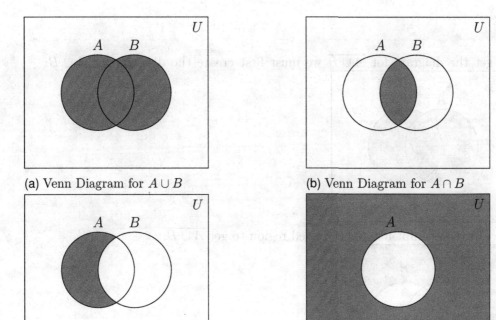

(a) Venn Diagram for $A \cup B$

(b) Venn Diagram for $A \cap B$

(c) Venn Diagram for $A - B$

(d) Venn Diagram for $\overline{A}$

Figure 1.1 Venn diagrams for set operations

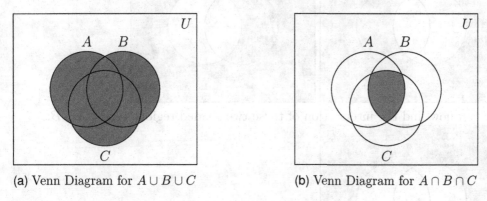

(a) Venn Diagram for $A \cup B \cup C$

(b) Venn Diagram for $A \cap B \cap C$

Figure 1.2 Venn diagrams for 3-way set operations

Example 1.30

To get the diagram for $\overline{A \cup B}$ we must first create the diagram for $A \cup B$:

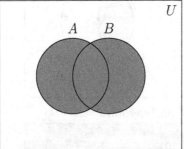

We can then complement the shaded region to get $\overline{A \cup B}$:

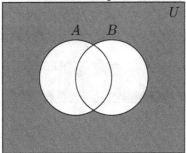

To get the diagram for $\overline{A} \cap \overline{B}$ we start with the diagrams for $\overline{A}$ and $\overline{B}$:

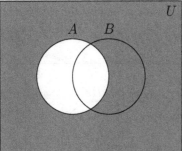 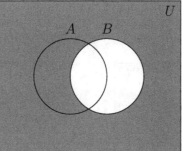

We can now find the intersection of these two shaded regions to get $\overline{A} \cap \overline{B}$:

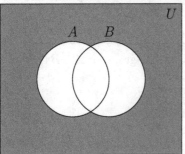

Notice that the diagrams for $\overline{A \cup B}$ and $\overline{A} \cap \overline{B}$ are identical, showing that the two expressions are equivalent.

Try and show that $\overline{A \cap B} = \overline{A} \cup \overline{B}$ using an approach similar to the above example.

1.7 POWER SETS

It is sometimes useful to refer to the set of all subsets of a set.

Definition 1.5

The **power set** of a set is the set of all subsets of that set (including $\emptyset$ and the set itself).

Example 1.31

The power set of the set $A = \{0, 1\}$ is written as $\mathcal{P}(A)$ and is equal to $\{\emptyset, \{0\}, \{1\}, \{0, 1\}\}$. Note that it is a set of four sets.

Example 1.32

The power set of the set $A = \{0, \{1\}\}$ is $\mathcal{P}(A) = \{\emptyset, \{0\}, \{\{1\}\}, \{0, \{1\}\}\}$. Note that A contains the set $\{1\}$ as an element so that the power set of A has elements sets that themselves contain sets.

We can also have even more complex power sets.

Example 1.33

The power set of the power set of $A = \{0, \{1\}\}$ is $\mathcal{P}(\mathcal{P}(A)) = \{\emptyset, \{\emptyset\}, \{\{0\}\}, \{\{\{1\}\}\}, \{\{0, \{1\}\}\}, \{\emptyset, \{0\}\}, \{\emptyset, \{\{1\}\}\}, \{\emptyset, \{0, \{1\}\}\}, \{\{0\}, \{\{1\}\}\}, \{\{0\}, \{0, \{1\}\}\}, \{\{\{1\}\}, \{0, \{1\}\}\}, \{\emptyset, \{0\}, \{\{1\}\}\}, \{\emptyset, \{0\}, \{0, \{1\}\}\}, \{\emptyset, \{\{1\}\}, \{0, \{1\}\}\}, \{\{0\}, \{\{1\}\}, \{0, \{1\}\}\}, \{\emptyset, \{0\}, \{\{1\}\}, \{0, \{1\}\}\}\}$. Note that the subsets have been listed in order of their cardinality—from the empty set of size zero to the entire power set of A of size 4. The easiest way to generate a set such as this is to give names to the subsets of A, say $B = \emptyset$, $C = \{0\}$, $D = \{\{1\}\}$, and $E = \{0, \{1\}\}$, then generate the power set of $\{B, C, D, E\}$ and, finally, replace B, C, D, and E with their original sets everywhere.

Since each element of a set can either be included or not included in a subset, we can have an exponential number of subsets. More specifically, the cardinality of the power set of a set A is given by

$$|\mathcal{P}(A)| = 2^{|A|}.$$

Example 1.34

The size of the power set of $A = \{1, 2, 3\}$ is $2^{|A|} = 2^3 = 8$. The size of the power set of $B = \{1, 2, 3, \ldots, 100\}$ is 2^{100}, which is a really large number (1267650600228229401496703205376).

1.8 TUPLES AND CARTESIAN PRODUCTS

Next, we look at the tuple type that is closest to the notion of a list in most programming languages.

Definition 1.6

A **tuple** is an ordered collection of items that allows for repeats.

Tuples are comma-separated lists with parentheses on both ends, such as $(1, 2, 3)$. They are different from sets in that they allow repeats, and the order of items actually matters.

Example 1.35

The tuples $(1, 2, 2)$ and $(1, 2)$ and $(2, 1, 2)$ are distinct. In contrast, the sets $\{1, 2, 2\}$ and $\{1, 2\}$ and $\{2, 1, 2\}$ are all the same set.

Tuples that have two items are called **pairs**. Tuples with three items are called **triples**. Longer tuples are referred to as n-**tuples**, where n is the length of the tuple.

Example 1.36

The tuple $(1, 2)$ is a pair. The tuple $(1, 2, 4)$ is a triple. The tuple $(1, 2, 4, 8, 16)$ is an example of a 5-tuple.

An important way in which tuples arise is when we look at all combinations of items in two or more sets, called the **Cartesian product**.

Example 1.37

If we want to look at all possible combinations of orders at a restaurant in which the main courses are $M = \{\text{burger, salad, sandwich}\}$ and the drink options are $D = \{\text{water, soda, juice}\}$, then the possible orders are denoted by the Cartesian product $M \times D$, which is the set of pairs $\{$(burger, water), (burger, soda), (burger, juice), (salad, water), (salad, soda), (salad, juice), (sandwich, water), (sandwich, soda), (sandwich, juice)$\}$.

The Cartesian product can be done in three (or more) ways as well, as seen in the following example:

Example 1.38

If $A = \{1, 2\}$, $B = \{3, 4\}$, and $C = \{5\}$, then $A \times B \times C$ is the set of triples $\{(1, 3, 5), (1, 4, 5), (2, 3, 5), (2, 4, 5)\}$.

Example 1.39

If $A = \{1, 2\}$, $B = \{3\}$, $C = \{4\}$, and $D = \{5\}$, then $A \times B \times C \times D$ is the set of 4-tuples $\{(1, 3, 4, 5), (2, 3, 4, 5)\}$.

You can denote the Cartesian product of a set with itself using power notation.

Example 1.40

If $A = \{a, b\}$ then $A^2 = A \times A = \{(a, a), (a, b), (b, a), (b, b)\}$. Similarly, $A^3 = \{(a, a, a), (a, a, b), (a, b, a), (a, b, b), (b, a, a), (b, a, b), (b, b, a), (b, b, b)\}$.

One has to be careful if parentheses are added to the Cartesian product.

Example 1.41

If $A = \{1, 2\}$, $B = \{3, 4\}$, and $C = \{5\}$, then $A \times (B \times C)$ is the set $\{(1, (3, 5)), (1, (4, 5)), (2, (3, 5)), (2, (4, 5))\}$. Notice that each element is a pair in which the second item of each pair is a pair.

The cardinality of a Cartesian product can be computed as the product of the cardinalities of the individual sets. If we have sets $A_1, A_2, A_3, \ldots, A_n$, then the cardinality of the Cartesian product $A_1 \times A_2 \times A_3 \times \ldots \times A_n$ is $|A_1| \cdot |A_2| \cdot |A_3| \cdot \ldots \cdot |A_n|$.

Example 1.42

Let $A = \{1, 2, 3, 4, 5, 6, 7, 8, 9, 10\}$ and $B = \{11, 12, 13\}$. Then the cardinality of $A \times B$ is $|A| \cdot |B| = 10 \cdot 3 = 30$. Similarly, the cardinality of $A \times A \times B$ is $|A| \cdot |A| \cdot |B| = 10 \cdot 10 \cdot 3 = 300$.

A special case to consider is when the Cartesian product involves a set that is empty. By the above cardinality relationships, we know that for any set A we have that $A \times \emptyset$ must have cardinality $|A| \cdot 0 = 0$ and must therefore be the empty set.

Another way to see this is to consider that the result of the Cartesian product can be computed by Algorithm 1.1. If any of the nested loops runs zero times, then the innermost line is never run, meaning that S remains empty.

Algorithm 1.1: CartesianProduct(A, B)

input : Two sets A and B
output: The Cartesian product $A \times B$
 1: let $S = \emptyset$
 2: **for all** $a \in A$ **do**
 3:    **for all** $b \in B$ **do**
 4:       insert (a, b) into S
 5: return S

1.9 FUNCTIONS

You will make extensive use of functions in computer science and this book.

> **Definition 1.7**
>
> **Functions** are mappings of values in one set (called the **domain**) to values in another set (called the **codomain**).

It is useful to think of the domain as the types of the input parameters to our function, as we would in a strongly typed language such as Java, and the codomain as the type of the output of the function.

> **Example 1.43**
>
> The function $f : \mathbb{Z} \to \mathbb{N}$ defined by $f(x) = x^2$ maps integers into the natural numbers. Here $\mathbb{Z}$ is the domain and $\mathbb{N}$ is the codomain.

> **Example 1.44**
>
> The function $FtoC : \mathbb{R} \to \mathbb{R}$ defined by $FtoC(f) = 5(f - 32)/9$ maps temperatures in degrees Fahrenheit to degrees Celsius.

Later in this book, we will see functions that have more complex domains and codomains over tuples.

> **Example 1.45**
>
> The function $add : \mathbb{R} \times \mathbb{R} \to \mathbb{R}$ defined by $add(x, y) = x + y$ has a domain $\mathbb{R} \times \mathbb{R}$, which we can think of as having an input that is a tuple.

> **Example 1.46**
>
> The function $swap : \mathbb{R} \times \mathbb{R} \to \mathbb{R} \times \mathbb{R}$ defined by $swap(x, y) = (y, x)$ has a domain and codomain of $\mathbb{R} \times \mathbb{R}$.

> **Example 1.47**
>
> The function $mult : \{1, 2, 3\} \times \{1, 2, 3, 4\} \to \mathbb{N}$ defined by $mult(x, y) = x \cdot y$ can be defined by the following table in which the first argument (x) is on the rows and the second argument (y) is on the columns:
>
> | | 1 | 2 | 3 | 4 |
> |---|---|---|---|---|
> | 1 | 1 | 2 | 3 | 4 |
> | 2 | 2 | 4 | 6 | 8 |
> | 3 | 3 | 6 | 9 | 12 |

1.10 STRINGS

Strings are mathematical objects that are very similar to strings in programming languages. They are sequences of characters such as abc or 123. Similar to programming languages, strings can be **concatenated** with one another so that if we have two strings $s_1 = $ abc and $s_2 = $ def then the concatenated string $s_1s_2 = $ abcdef.

Strings are defined over an **alphabet** set Σ (the Greek symbol uppercase sigma).

Example 1.48

Some strings from the alphabet $\Sigma = \{$A, C, G, T$\}$ include GATTACA and ATAT.

Notice that, unlike programming language strings, we do not include any quotes around the string. Another important difference is that programming languages allow for all characters (ASCII or Unicode), but mathematical strings will usually be restricted to a smaller collection of characters.

Definition 1.8

A **string** is a finite, ordered collection of characters from a finite alphabet Σ.

The **length** of a string s is the number of characters in it and denoted $|s|$.

Example 1.49

The length of the string GATTACA is 7, written as $|$GATTACA$| = 7$ and the length of the string ATAT is 4 ($|$ATAT$| = 4$).

The unique string of length zero is called the **empty string**. It is usually denoted in programming languages by "", but since no quotation marks are used for mathematical strings, we use the Greek symbol lowercase lambda (λ). Some books also use lowercase epsilon (ϵ) to denote the empty string.

A shorthand that we will use for strings such as aaaaaaa is to use power notation to express it more succinctly as a^7. We can also combine this with concatenation to write strings such as $a^3b^2a^3$ to be shorthand for aaabbaaa. This shorthand can also be applied to string expressions, such as $(ab)^3$ for ababab.

A **substring** is any consecutive sequence of characters in a string.

Example 1.50

The substrings of the string abc are λ, a, b, c, ab, bc, and abc.

A **prefix** is any initial substring of a string.

Example 1.51

The prefixes of the string abc are λ, a, ab, and abc.

A **suffix** is any substring at the end of a string.

The suffixes of the string abc are λ, c, bc, and abc.

Notice that strings above were listed in ascending order in their length, breaking ties using the usual dictionary order. This is known as **shortlex** ordering.

In shortlex ordering, a string s_1 appears before s_2 ($s_1 < s_2$)

1. if s_1 is shorter than s_2; or

2. if they are of the same length, the first character at which s_1 differs from s_2 comes earlier in the alphabet.

This is easiest seen by the way of a few examples:

Over $\Sigma = \{a, b, c\}$, the string abc comes earlier than aabc as the former is shorter than the latter. Similarly, cc comes before aaa.

Note that shortlex ordering is not the same as standard dictionary ordering as, for example, the string cat comes before the string abacus (since cat is shorter).

Over $\Sigma = \{a, b, c\}$, the string aac comes earlier than abc because they are the same length and the first place that they differ (the second position) has an a in the first and b for the second. Similarly, cabc comes before caca.

We will refer to all the strings over an alphabet Σ as Σ^*. Using the notion of shortlex ordering, we can now write all strings in Σ^* in the following example:

Over $\Sigma = \{0, 1\}$, we can order Σ^* as follows: $\lambda, 0, 1, 00, 01, 10, 11, 000, 001, \ldots$.
Over $\Sigma = \{a, b, c\}$, we can order Σ^* as follows: λ, a, b, c, aa, ab, ac, ba, bb, bc, ca, cb, cc, aaa, ….

1.11 LANGUAGES

Almost all the computation we will see in this book will consist of recognizing or accepting languages.

Definition 1.10

A **language** over Σ is a subset of Σ^*.

Example 1.56

Over $\Sigma = \{0,1\}$, the language $\{s : s$ has length $2\}$ is $\{00, 01, 10, 11\}$.

Example 1.57

Over $\Sigma = \{0,1\}$, the language

$$\{s : s \text{ starts with a } 0\}$$

is $\{0, 00, 01, 000, 001, 010, 011, 0000, \ldots\}$.

Example 1.58

Over $\Sigma = \{0,1\}$, the language $\{s : s$ has an even number of 1s and no 0s$\}$ is $\{\lambda, 11, 1111, 111111, \ldots\}$. Note that the empty string is included because it has zero 1s and zero is considered an even number.

We can write out languages more succinctly and precisely by using mathematical notation.

Example 1.59

We can rewrite $\{s \in \{0,1\}^* : s$ has an even number of 1s and no 0s$\}$ as $\{1^{2n} : n \in \mathbb{N}\}$. Note that the $2n$ exponent will generate precisely the even natural numbers and that 1^0 is the empty string (since we repeat 1 zero times). We thus generate precisely the same set of strings $\{\lambda, 11, 1111, 111111, \ldots\}$

Example 1.60

We can rewrite

$$\{s \in \{0,1\}^* : s \text{ has an even number of 1s followed by an odd number of 0s}\}$$

as $\{1^{2n}0^{2m+1} : n, m \in \mathbb{N}\}$. Notice this time that the expression $2m + 1$ will generate all the odd natural numbers. It is also important to understand why two different variables (n and m) were used. If we used n for both exponents, then we would only generate strings that have one more 0 than 1 and would not generate strings such as 1100000.

We can also introduce string variables to write out sets over strings more compactly, as shown in the following examples:

Example 1.61

The set of strings over $\{0,1\}$ that start with a 0 can be written as:

$$\{0s : s \in \{0,1\}^*\}.$$

Here, the string variable s can be any string over $\{0,1\}$, including the empty string, so we get the set $\{0, 00, 01, 000, 001, \ldots\}$.

Example 1.62

The set of strings over $\{0,1\}$ that contain the substring 010 can be written as:

$$\{x010y : x, y \in \{0,1\}^*\}.$$

Here x and y represent all possible prefixes and suffixes of a string of the given form, including the empty string.

The term language comes from the fact that the languages that you are familiar with can be represented by sets of strings.

Example 1.63

We can define the natural language

$$\text{ENGLISHSENTENCES} = \{s : s \text{ is a grammatical English sentence}\}$$

as a set over all the letters and punctuation symbols in English. In fact, in the CS sub-discipline of Natural Language Processing (NLP), a goal is to parse sentences so as to recognize and interpret them.

Example 1.64

We can define programming languages as all valid programs, such as

$$\text{PYTHON} = \{s : s \text{ is a valid Python program}\}.$$

One of the first tasks that a programming language interpreter or compiler does is to parse your program to make sure that it is syntactically valid.

The computational tasks we will perform in this book then just become ones of determining whether a string (the input) is in a particular language. Any function that we can compute that has a Boolean output (True or False) can be thought of as the set of input strings that return True.

Example 1.65

When computing whether a number is prime or not, we are effectively determining membership in the language

$$\{s \in \{0, 1, 2, 3, 4, 5, 6, 7, 8, 9\}^* : s \text{ is a prime number}\}.$$

This may appear unusual to you as computations such as addition and multiplication (or more complex functions) aren't tasks where you are trying to determine if something is in a set. While it is true that we lose the ability to describe such computation with languages, we still retain considerable expressive power that is sufficient to prove many results about various models of computation.

1.12 CHAPTER SUMMARY AND KEY CONCEPTS

- A **set** is an unordered collection of distinct objects (numbers, strings, programming languages, etc.). The objects in a set are called its **elements** or **members**. The notation $x \in A$ is how we denote that x is an element of the set A and $x \notin A$ is how we denote that x is not an element of the set A.

- The **empty set** ($\emptyset$) is the unique set with no elements.

- A set is a **subset** of another set if all its elements are contained in the other set. The empty set is a subset of all sets. A set is always a subset of itself.

- Two sets are **equal** if they have the same elements. To show that two sets are equal, we have to show that both sets are subsets of each other.

- The **cardinality** of a set is the number of elements in the set. Sets can be **finite** or **infinite**.

- The set $\mathbb{N}$ is all the **natural numbers**, the set $\mathbb{Z}$ is the set of **integers**, the set $\mathbb{R}$ is the set of all **real numbers**, and the set $\mathbb{Q}$ is the set of all **rational numbers** (i.e., numbers of the form p/q, where p and q are integers and q is not zero).

- Sets are often expressed in **set builder notation**, which takes the form { expression : constraints}.

- The **union** of sets A and B is $A \cup B = \{x : x \in A \text{ or } x \in B\}$. The **intersection** of sets A and B is $A \cap B = \{x : x \in A \text{ and } x \in B\}$. The **difference** of sets A and B is $A - B = \{x : x \in A \text{ and } x \notin B\}$.

- The **complement** of a set A is its difference from a universal set U, $\overline{A} = U - A$.

- **Venn diagrams** are diagrams that visualize set operations.

- The **power set** of a set is the set of all its subsets. The cardinality of the power set of a set A is given by $|\mathcal{P}(A)| = 2^{|A|}$.

- A **tuple** is an ordered collection of items that allows for repeated items. Sets of tuples are commonly created by computing **Cartesian products** of sets. The size of the set created from the Cartesian product of several sets is the product of the sizes of those sets.

- **Functions** are mappings of values from a **domain** set to a **codomain** set. Functions can be represented by formulas or tables.

- A **string** is an ordered sequence of characters from an **alphabet**. The number of characters in a string s is called its **length** and is denoted by $|s|$. The unique string of length zero is called the **empty string** λ.

- Any consecutive sequence of characters in a string is called a **substring**. When the substring appears at the start of the string, it is called a **prefix**. When it appears at the end of the string, it is called a **suffix**.

- Strings can be sorted using **shortlex ordering**. In this ordering, shorter strings come before longer strings. For strings of the same length, the dictionary ordering is used—the leftmost character at which the two strings differ is used to order them.

- Any set of strings over an **alphabet** Σ is called a **language**. Languages are used extensively in computer science to define sets with the same property (e.g., whether a number is a prime or whether a computer program is syntactically correct).

EXERCISES

Exercises for 1.4

Write out the elements of the following sets within set braces:

1.1 The set of programming languages you have coded in

1.2 The set of the days of the week

1.3 $\{2n : n \in \mathbb{N}, n < 5\}$

1.4 $\{3n : n \in \mathbb{N}, 7 \leq n \leq 10\}$

1.5 $\{7n : n \in \mathbb{N}\}$

1.6 $\{10n : n \in \mathbb{Z}\}$

1.7 $\{4n + 1 : n \in \mathbb{N}\}$

1.8 $\{3n - 1 : n \in \mathbb{Z}\}$

1.9 $\{2n^2 : n \in \mathbb{Z}\}$

1.10 $\{n^3 : n \in \mathbb{N}\}$

Write the following sets using set-builder notation:

1.11 $\{4, 5, 6, \ldots, 40\}$

1.12 $\{-3, -2, -1, \ldots, 17\}$

1.13 $\{0, 2, 4, 6, \ldots\}$

1.14 $\{0, 1, 4, 9, 16, \ldots\}$

1.15 $\{3, 7, 11, 15, 19, \ldots\}$

Exercises for 1.5

Assuming that $A = \{1, 2, 3, 4, 5\}$, $B = \{2, 3, 5, 7\}$, $C = \{4, 5, 6, 7, 8\}$ and a universal set $U = \{1, 2, 3, 4, 5, 6, 7, 8\}$, compute the following:

1.16 $A \cup B$

1.17 $B \cap C$

1.18 $A - B$

1.19 $B - C$

1.20 $\overline{C}$

1.21 $(A \cap B) \cup C$

1.22 $\overline{A \cap B}$

1.23 $(A \cup B) \cap (\overline{B \cup C})$

Exercises for 1.6

Draw Venn diagrams for the following expressions, showing the steps (intermediate diagrams) in each case:

1.24 $(A - B) \cup (B - A)$

1.25 $(A \cap B) \cup \overline{A}$

1.26 $(A \cup B) - C$

1.27 $\overline{A} \cup \overline{B} \cup \overline{C}$

1.28 $\overline{A} \cap \overline{B} \cap \overline{C}$

Show using Venn diagrams, using labeled intermediate steps, whether the following pairs of expressions are equal:

1.29 $A \cup (B \cap C)$ and $(A \cap B) \cup (A \cap C)$

1.30 $A \cup (B \cap C)$ and $(A \cup B) \cap (A \cup C)$

1.31 $C - (A \cap B)$ and $(C - A) \cup (C - B)$

1.32 $C - (A \cup B)$ and $(C - A) \cap (C - B)$

1.33 $A - B$ and $A \cap \overline{B}$

Exercises for 1.7

List the elements of the following sets within set braces:

1.34 $\mathcal{P}(\{0, 1\})$

1.35 $\mathcal{P}(\{a, b, c\})$

1.36 $\mathcal{P}(\{a, \{b, c\}\})$

1.37 $\mathcal{P}(\mathcal{P}(\{0, 1\}))$

1.38 $\mathcal{P}(\mathcal{P}(\mathcal{P}(\{0\})))$

Compute the *size* of each of the following sets:

1.39 $\mathcal{P}(\{0, 1, 2, 3, 4\})$

1.40 $\mathcal{P}(\{0, 1, \{2, 3, 4\}\})$

1.41 $\mathcal{P}(\{\{0, 1\}, \{2, 3, 4\}\})$

1.42 $\mathcal{P}(\{\{0, 1\}, \{2, 3\}, 4\})$

1.43 $\mathcal{P}(\mathcal{P}(\{0, 1, 2, 3, 4\}))$

1.44 $\mathcal{P}(\mathcal{P}(\mathcal{P}(\{0, 1\})))$

Exercises for 1.8

Assuming that $A = \{0, 1\}, B = \{a, b\}$ and $C = \{1, 2, 3\}$, list the elements of the following sets within set braces:

1.45 $A \times B$

1.46 $B \times C$

1.47 C^2

1.48 A^4

1.49 $A \times B \times B$

1.50 $A \times B \times C$

1.51 $(A \times B) \times C$

1.52 $A \times (B \times C)$

Assuming that $A = \{0, 1\}, B = \{a, b\}$ and $C = \{1, 2, 3\}$, compute the *size* of each of the following sets:

1.53 $A \times C$

1.54 $A \times B \times A \times B$

1.55 A^5

1.56 C^{10}

1.57 $\mathcal{P}(A) \times \mathcal{P}(B)$

1.58 $\mathcal{P}(\mathcal{P}(A) \times \mathcal{P}(B) \times \mathcal{P}(C))$

Exercises for 1.9

List the domain and the codomains for the following functions:

1.59 A function `double` that takes as input a real number and returns twice its value

1.60 A function `max` that takes as input 5 integers and returns the largest one

1.61 A function `sum` that takes as input 10 real numbers and returns their sum

1.62 A function `sort5` that takes as input 5 integers and returns them sorted

Exercises for 1.10

Write out the following strings in shorthand notation:

1.63 aaaaaaa

1.64 aaabbbbccc

1.65 aaaaabbaaaaa

1.66 abcabcabcabc

1.67 aaaaababababababccccc

Write out all the prefixes and suffixes for the following strings:

1.68 aaaa

1.69 ababa

1.70 aaaaabbbbbb

1.71 aaabbccc

1.72 abcabcabc

Exercises for 1.11

List all strings of length at most 4 from the following languages:

$$\Sigma = \{0, 1\}$$

1.73 $\{0^n : n \in \mathbb{N}\}$

1.74 $\{0^n 1^n : n \in \mathbb{N}\}$

1.75 $\{0^n 1^m : n, m \in \mathbb{N}\}$

1.76 $\{0^n 1^m : n, m \in \mathbb{N}, n + m \geq 2\}$

1.77 $\{x01 : x \in \Sigma^*\}$

1.78 $\{x11y : x, y \in \Sigma^*\}$

Write the following languages in set builder (mathematical) notation:

$$\Sigma = \{0, 1\}$$

1.79 The set of all strings which consist of zero or more 0s and no 1s.

1.80 The set of all strings which consist of one or more 1s and no 0s.

1.81 The set of all strings which consist of one or more 0s followed by one or more 1s.

1.82 The set of all strings that consist of a number of 0s followed by the same number of 1s (at least one of both).

1.83 The set of all strings that consist of a number of 0s followed by a larger number of 1s (at least one of both).

1.84 The set of all strings with a prime number of 1s and no 0s.

$$\Sigma = \{a, b\}$$

1.85 The set of all strings that begin with an **a**.

1.86 The set of all strings that begin with an **a** and end with a **b**.

1.87 The set of all strings that have an even number of **a**s and no **b**s.

1.88 The set of all strings that have an even number of **a**s followed by an odd number of **b**s.

1.89 The set of all strings that contain **ab** as a substring.

1.90 The set of all strings that have an **a** in the third position.

1.91 The set of all strings that have a **b** four positions from the end.

1.92 The set of all strings that start and end with the same three characters.

CHAPTER 2

Deterministic Finite Automata

2.1 WHY YOU SHOULD CARE

We are now ready to study a simple model of computation. As computer scientists, we have to understand the computing models that we working with to determine what is easily computable and (later in this book) what is not computable at all. The first model of computation that we will study is called a **deterministic finite automaton** or **DFA** (plural: deterministic finite automata).

2.2 A VENDING MACHINE EXAMPLE

Suppose that you had a vending machine that only accepts nickels (5 cent coins) and dimes (10 cent coins) and sold snacks for 20 cents. (Wouldn't it be lovely to be able to buy anything for 20 cents!) The DFA for such a machine is given in this example.

> **Example 2.1**
>
> DFA for a vending machine
>
>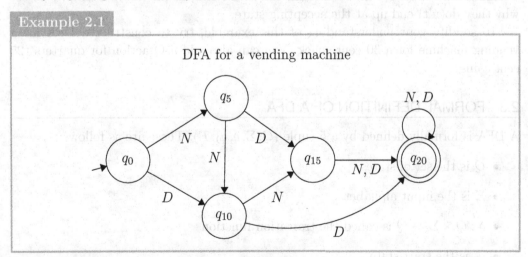

The circles in the diagram are called states and the computation is always at exactly one state at any given time. Any computation on this DFA starts at the start state (q_0 in this example) that is denoted by an arrow entering it. The actions that are then performed are the insertion of nickels (denoted with an arrow labeled with an N) or dimes (denoted by arrows labeled with a D). These actions cause the computation to transition between the states $\{q_0, q_5, q_{10}, q_{15}, q_{20}\}$. Note that arrows with multiple characters have commas separating them, for example we can transition from q_{15} to q_{20} on either N or D.

The names of the states were picked in such a way to denote how much money has been entered into the vending machine. For example, when a dime is entered at state q_0, the vending machine should transition to the state q_{10} denoting a total of 10 cents that have been entered thus far.

The state q_{20} is called the final or accepting state and indicates the condition under which the vending machine dispenses a snack. It is denoted by a double circle.

For a DFA to be completely defined, each state must have a transition for each possible input (in this case a nickel or dime). In particular, in this example, we send any over-payment (a dime at q_{15} or any additional coins at q_{20}) to q_{20}. Conceptually, this represents the idea that the machine should still dispense the snack if there is over-payment—it may just eat change.

Rather than performing any actions, this DFA is designed to capture the set of inputs that should cause the machine to dispense a snack. For example, if a user enters a nickel and a dime, this input would be represented by the string ND and since it would end at state q_{15} (a non-final state), this input would be rejected—there was insufficient money for the snack.

Example 2.2

The vending machine accepts the inputs NDN, DD, DDDD, NDND, and DND. The vending machine rejects the inputs NN, ND, NNN, and DN.

Make sure to follow the transitions for each string in the above examples to see why they do(n't) end up at the accepting state.

To solidify your understanding of this example, try to construct a DFA for a vending machine for a 30 cent snack. You can also add a Q action for quarters (25 cent coins).

2.3 FORMAL DEFINITION OF A DFA

A DFA is formally defined by a 5-tuple $(Q, \Sigma, \delta, q_0, F)$. These are as follows:

- Q is the set of states
- Σ is the input alphabet
- $\delta : Q \times \Sigma \to Q$ is called the transition function
- q_0 is the start state
- F is the set of final or accepting states

These are best illustrated with examples.

Example 2.3

The DFA for the vending machine example is given by the tuple $(Q, \Sigma, \delta, q_0, F)$, where

- $Q = \{q_0, q_5, q_{10}, q_{15}, q_{20}\}$

- $\Sigma = \{N, D\}$

- δ is given by the table:

	N	D
q_0	q_5	q_{10}
q_5	q_{10}	q_{15}
q_{10}	q_{15}	q_{20}
q_{15}	q_{20}	q_{20}
q_{20}	q_{20}	q_{20}

- $F = \{q_{20}\}$

When designing your own DFAs, you should make sure that the table for the δ transition function is fully filled in—every pair of state and input symbol must have an entry. This is something easy to forget when you are first learning to create DFAs.

Notice how the entire computation model can be so precisely captured using the mathematical data types that you learned about in Chapter 1, including sets, tuples, alphabets, strings, and functions. This is conceptually similar to "coding" up your computation using technical notation.

We give a name to all the languages accepted by a DFA:

Definition 2.1

A **regular language** is one that has a DFA that accepts it.

2.4 MATCHING PHONE NUMBERS

The next example of a DFA we will look at is for one that will identify numbers that start with the author's area code (740). This can be represented by the language

$$\{740s : s \in \{0, \ldots, 9\}^*\}.$$

The next figure shows a DFA that will accept only strings that start with 740. (Note that it will accept strings of any length that start with 740, including ones that are too short or long to be a phone number, but we will ignore that detail for now.)

Example 2.4

DFA for phone number problem

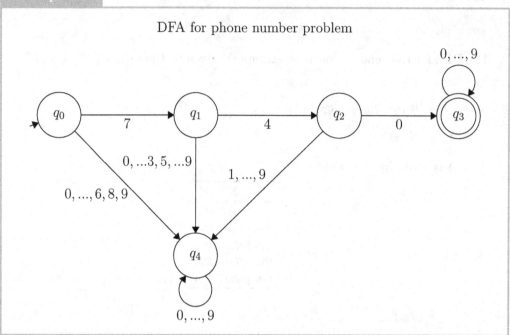

This DFA works by moving toward the accepting state (q_3) precisely when the first three symbols are 7, 4, and 0. Notice that the states q_3 and q_4 are what are known as **sink states**. That is, it is impossible to leave them once entered. The difference, of course, is that once the state q_3 is entered, the string is accepted no matter what (it has been confirmed to start with 740), whereas once the state q_4 is entered the string is definitely rejected since it did not start with a 740 pattern.

As before, we can create a tuple for this DFA.

Example 2.5

The phone number DFA is given by the tuple $(Q, \Sigma, \delta, q_0, F)$, where

- $Q = \{q_0, q_1, q_2, q_3, q_4\}$

- $\Sigma = \{0, \ldots, 9\}$

- δ is given by the table:

	0	1	2	3	4	5	6	7	8	9
q_0	q_4	q_4	q_4	q_4	q_4	q_4	q_4	q_1	q_4	q_4
q_1	q_4	q_4	q_4	q_4	q_2	q_4	q_4	q_4	q_4	q_4
q_2	q_3	q_4	q_4	q_4	q_4	q_4	q_4	q_4	q_4	q_4
q_3	q_3	q_3	q_3	q_3	q_3	q_3	q_3	q_3	q_3	q_3
q_4	q_4	q_4	q_4	q_4	q_4	q_4	q_4	q_4	q_4	q_4

- $F = \{q_3\}$.

A common error to avoid with DFAs is to be imprecise about its language. In particular, someone looking at this DFA might notice that it accepts all strings that start with 74 and infer that that is its language. Of course, since this DFA does not accept strings such as 747, this is not correct—saying that it accepts all strings that start with 74 was too broad.

The opposite problem might occur as well. It might be noted that this DFA does accept all strings that start with 7405. This time, the observation is too specific—this characterization excludes strings such as 7401234567. Being precise about the language of a DFA you are given or that you design is very important. Just as with writing code, you want to test your DFA with lots of examples and think about special cases.

2.5 COMPUTATIONAL BIOLOGY

You might have heard that DNA is encoded by four types of nucleotides that are represented by the letters A, C, G, and T. Sequences of these nucleotides are studied by computational biologists to analyze DNA and search for proteins. Our next problem will be to take a string of DNA and determine whether the substring of length three (called a codon) TTA appears anywhere in it. This can be written as the language

$$\{pTTAs : p, s \in \{A, C, G, T\}^*\}.$$

For example, the string GATTACA should be accepted but the string CATCAT should not. A DFA for this language is given in the example below:

Example 2.6

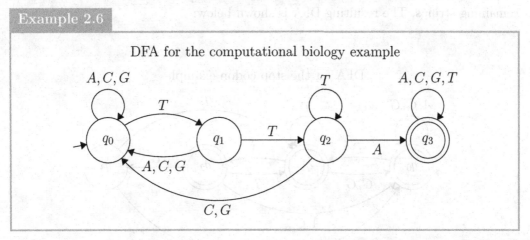

DFA for the computational biology example

Similar to the previous example, we create a path to the accepting state when TTA is seen as a substring ($q_0 \rightarrow q_1 \rightarrow q_2 \rightarrow q_3$). This time, however, the substring can be in any part of the string, so we have to be careful about the other transitions. It is useful here to think about what strings should reach each of the states. Since we have to keep track of whether the substring TTA has appeared, we use states to remember our progress toward this goal. We use q_1 to denote the case that we have not yet found a TTA substring and we have most recently encountered a single T (but not a TT). The state q_2 is reachable by strings for which we have not yet found a TTA substring and we have most recently encountered a TT. The accepting sink state q_3 is reached once we have encountered the entire substring TTA. The start state q_0 then is reachable from all other strings.

Armed with this intuition, we can understand why all the transitions were chosen in Example 2.6. Any transition out of a state that does not make progress toward the goal of identifying the string TTA should go back to state q_0. For example, the state q_1 returns to q_0 on an A, C, or G. The only exception here is that q_2 on a T self-loops since the past two characters were still TT, so we want to stay on q_2 on a T.

As with the other DFAs, trace the path for strings in the language such as TTA, ATTGTTA, and ATTTTAGC to see why they get to the accepting state as well as why other strings such as TTG, TATTCTT, and TTTTGTT do not get to the accepting state.

Generating the tuple for this DFA is left as an exercise for the reader.

2.6 STOP CODONS

Our next example, also from computational biology, will be to identify strings that do *not* end with the stop codon TAG. That is, we want to accept the complement of the language $\{xTAG : x \in \{A, C, G, T\}^*\}$. Just like the previous example, we need to create states that will allow us to keep track of whether the last three symbols were T, A, and G. We will use a state q_1 to represent whether the last symbol seen was a T, q_2 to represent whether the last two symbols were TA, and q_3 to represent whether the last three symbols were TAG. The state q_0 will once again catch all the remaining strings. The resulting DFA is shown below:

Example 2.7

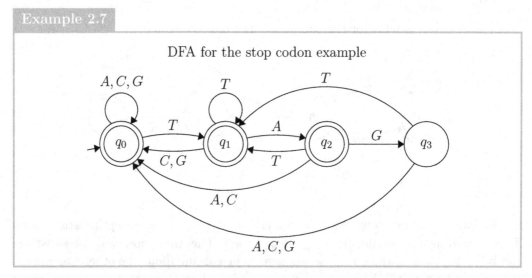

DFA for the stop codon example

Notice here that, unlike the previous example, we accept strings that reach any state other than q_3. Since we have multiple accepting states, F really has to be a set to hold all of them.

Again, practice creating the tuple for this DFA.

2.7   DIVVYING UP CANDY

A mom wants to split up the candy between her two kids and wants to know if she can divide it evenly between them. There are three types of candies: chocolates, lollipops, and taffy. The mom wants to make sure that there is an even number of chocolates and lollipops, but she doesn't care about the taffy since neither kid likes it.

We will create a DFA that will determine whether a pile of candies can be evenly divided two ways. The pile will be represented by a string in which chocolates, lollipops, and taffy are represented by the letters C, L, and T, respectively. For example, the DFA should accept the strings CLTTLLLC and TCCTT since they have an even number of chocolates and an even number of lollipops. The DFA should reject the string TLLC and TLC because the first has an odd number of chocolates and the second has an odd number of both chocolates and lollipops.

To design a DFA for this problem, we have to think about the possible states that we need to represent. Specifically, we care if there is an even or an odd number of chocolates, and similarly for the lollipops. Since the number of chocolates and lollipops are unrelated numbers, we will have a total of four possibilities: both even, odd chocolates and even lollipops, even chocolates and odd lollipops, and both odd. Armed with this intuition, we design the DFA seen in the example below:

Example 2.8

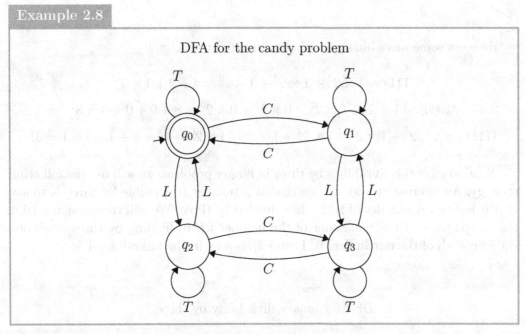

DFA for the candy problem

The states q_0, q_1, q_2, q_3 represent the four possibilities listed above. Each time we see another piece of chocolate we switch between the left and right states. Each time we see another lollipop we switch between and the top and bottom states. Since the number of pieces of taffy doesn't affect anything, we just self-loop on the current state when we see a taffy. Lastly, we make q_0 the accepting state since this is the one that captures strings that have an even number of chocolates and lollipops—a nice even split for the kids!

Once again, write out the full tuple for this DFA as practice.

2.8 DIVISIBILITY IN BINARY

Finite automata can also be used to compute mathematical properties. For example, we will see how to determine whether a number, given in binary, is divisible by three next. Before we do this, we will take a short detour to review binary numbers.

In the decimal number system (or base 10) that we are used to using, the digit at each position indicates the corresponding power of ten's place for the number. For example, the decimal number 1234 can be written as

$$1234 = 1 \times 10^3 + 2 \times 10^2 + 3 \times 10^1 + 4 \times 10^0.$$

In other words, there is a 1 in the thousand's (10^3) place, a 2 in the hundred's (10^2) place, a 3 in ten's (10^1) place, and a 4 in the one's (10^0) place.

Similarly, binary (or base 2) numbers have places for the powers of two. For example, the binary number 1011_2 (here, the subscript 2 indicates that the number is in binary) can be written as

$$1011_2 = 1 \times 2^3 + 0 \times 2^2 + 1 \times 2^1 + 1 \times 2^0 = 8 + 0 + 2 + 1,$$

which is just the number eleven.

Example 2.9

Here are some more binary numbers converted to decimal:

$$111_2 = 1 \times 2^2 + 1 \times 2^1 + 1 \times 2^0 = 4 + 2 + 1 = 7$$
$$1000_2 = 1 \times 2^3 + 0 \times 2^2 + 0 \times 2^1 + 0 \times 2^0 = 8 + 0 + 0 + 0 = 8$$
$$11111_2 = 1 \times 2^4 + 1 \times 2^3 + 1 \times 2^2 + 1 \times 2^1 + 1 \times 2^0 = 16 + 8 + 4 + 2 + 1 = 31$$

Now, to solve the divisibility by three in binary problem, we will use the following strategy. An alternative way of saying that a number is divisible by three is to say that it leaves a remainder of zero when divided by three. We will thus design a DFA that keeps track of the remainder of the number when dividing by three, with one state for each of the remainders 0, 1, and 2, as seen in the following DFA:

Example 2.10

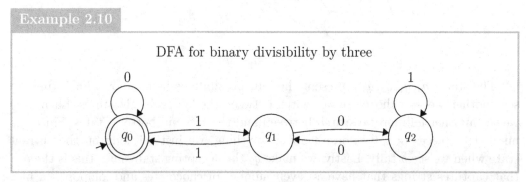

DFA for binary divisibility by three

To understand how the DFA was designed, it is first necessary to understand what happens to a binary number when a 0 or a 1 is appended to the end of it. When we add a 0 to the end of a binary number, it is the same as multiplying the number by 2 (similar to how adding a 0 to the end of a decimal number multiplies it by 10). When we add a 1 to the end of a binary number, it is the same as multiplying it by 2 and then adding 1. Therefore, when we start at state q_0 (the state representing a remainder of 0 when dividing by 3), doubling the number still gives a remainder of 0 when dividing by 3 so we have a self-loop on 0. On the other hand, when multiplying a number with remainder 0 by 2 and adding 1, we get a number that has a remainder of 1 when dividing by 3, so we transition from q_0 to q_1 on a 1. See what happens when you do the same for numbers that have a remainder of 1 when divided by 3 (state q_1) and numbers that have a remainder of 2 when divided by 3 (state q_2) to make sure you understand how the transitions out of q_1 and q_2 were determined.

Note that, for simplicity, we assume that the empty string is in the language. We also allow the binary number to have leading 0s, that is, we treat the number 0011_2 the same as the number 11_2.

2.9 CHAPTER SUMMARY AND KEY CONCEPTS

- A **deterministic finite automaton (DFA)** is a simple model for computing whether a string is in a language. It consists of **states**, including one **starting state** and one or more **accepting states**, with transitions between them.

- DFAs are formally defined by a **5-tuple** $(Q, \Sigma, \delta, q_0, F)$, where Q is the set of **states**, Σ is the **input alphabet**, δ is the **transition function**, q_0 is the **start state**, and F is the set of **final or accepting states**.

- A language is called **regular** if it has a DFA that accepts it.

- DFAs can be used to accept the languages with applications such as vending machines, pattern matching, computational biology, and testing arithmetic properties.

EXERCISES

Give the full 5-tuple for DFAs for the following languages:

2.1 $\{w \in \{0,1\}^* : w \text{ starts with a } 0\}$

2.2 $\{w \in \{0,1\}^* : w \text{ ends with a } 1\}$

2.3 $\{w \in \{0,1\}^* : w \text{ ends with } 00\}$

2.4 $\{w \in \{0,1\}^* : w \text{ has } 01 \text{ as a substring}\}$

2.5 $\{w \in \{0,1\}^* : w \text{ does not have } 01 \text{ as a substring}\}$

Draw DFAs for the following languages:

2.6 $\{w \in \{0,1\}^* : w \text{ is of odd length}\}$

2.7 $\{w \in \{0,1\}^* : w \text{ has an even number of } 0s\}$

2.8 $\{w \in \{0,1\}^* : w \text{ has an even number of 0s and an odd number of 1s}\}$

2.9 $\{w \in \{0,1\}^* : w \text{ has an even number of 0s and an even number of 1s}\}$

2.10 $\{w \in \{0,1\}^* : w \text{ has an odd number of 0s and an even number of 1s}\}$

2.11 $\{w \in \{0,1\}^* : w \text{ has an odd number of 0s and an odd number of 1s}\}$

2.12 $\{w \in \{0,1\}^* : \text{the third character in } w \text{ is } 0\}$

2.13 $\{w \in \{0,1\}^* : w \text{ starts and ends with a } 0\}$

2.14 $\{w \in \{0,1\}^* : w \text{ starts or ends with a } 0\}$

2.15 $\{w \in \{0,1\}^* : w \text{ has exactly one } 0\}$

2.16 $\{w \in \{0,1\}^* : w \text{ contains } 10 \text{ but not } 1010\}$

2.17 $\{w \in \{0,1\}^* : w \text{ contains at least two 0s and at least one 1}\}$

2.18 $\{w \in \{0,1\}^* : w \text{ contains at most two 0s and at least one 1}\}$

2.19 $\{w \in \{0,1\}^* : \text{the number of 0s in } w \text{ is divisible by 3}\}$

2.20 $\{w \in \{0,1\}^* : w \text{ is divisible by 4 in binary}\}$ (Let λ be in the language and ignore leading 0s.)

2.21 $\{w \in \{0,1\}^* : w \text{ is divisible by 5 in binary}\}$ (Let λ be in the language and ignore leading 0s.)

2.22 $\{w \in \{0,1\}^* : w \text{ has remainder 1 when divided by 5 in binary}\}$ (ignore leading 0s)

2.23 $\{w \in \{0,\ldots,9\}^* : w \text{ is even in decimal}\}$ (Let λ be in the language and ignore leading 0s.)

2.24 $\{w \in \{0,\ldots,9\}^* : w \text{ is divisible by 5 in decimal}\}$ (Let λ be in the language and ignore leading 0s.)

2.25 $\{w \in \{0,\ldots,9\}^* : w \text{ is divisible by 7 in decimal}\}$ (Let λ be in the language and ignore leading 0s.)

2.26 $\{w \in \{A,C,G,T\}^* : w \text{ has the substring ATG followed later by the substring TAA}\}$

2.27 $\{w \in \{A,C,G,T\}^* : w \text{ has the substring ATG followed later by the substring TAA or TAG or TGA}\}$

2.28 $\{w \in \{a,\ldots,z,0,\ldots,9\}^* : w \text{ is a password (string) of length at least 6 characters}\}$

2.29 $\{w \in \{a,\ldots,z,0,\ldots,9\}^* : w \text{ is a password (string) of length at least 6 characters with at least one digit}\}$

2.30 $\{w \in \{a, \ldots, z, 0, \ldots, 9\}^* : w$ is a password (string) of length at least 6 characters with at least one digit and one letter$\}$

2.31 $\{w \in \{0, 1\}^* :$ the third character from the end in w is 0$\}$

2.32 $\{w \in \{0, 1\}^* :$ at least one of the last three characters from the end is a 0$\}$

2.33 $\{w \in \{0, 1\}^* : w$ starts and ends with the same two characters$\}$

CHAPTER 3

Logic

3.1 WHY YOU SHOULD CARE

It is not an exaggeration to say that mathematical logic underlies all of computer science. Computer circuits are made out of logical operators. The mathematical arguments we make about algorithms follow the rules of logic. Moreover, understanding logic will make you better at writing conditional (`if`) statements and (`while`) loops when you code.

3.2 LOGICAL STATEMENTS

A **logical statement** is a mathematical statement that has a value of **true** or **false**.

Example 3.1

Here are some true statements:

- $2 + 2 = 4$
- $2 < 4$
- $1 \in \{1, 2, 3\}$
- $4 \notin \{1, 2, 3\}$

Example 3.2

Here are some false statements:

- $2 + 2 = 5$
- $2 > 4$
- $1 \notin \{1, 2, 3\}$
- $4 \in \{1, 2, 3\}$

DOI: 10.1201/9781003383284-3

We can treat logical statements like mathematical objects (for instance, sets or strings) and give them names.

> **Example 3.3**
>
> Let P be the statement "For every set S, $\emptyset \subseteq S$."
> Let Q be the statement "For every set S, $\emptyset \in \mathcal{P}(S)$."
> Let R be t he statement "For every set S, $S \cup \{1\} = S$."
> The statements P and Q are true, while the statement R is false (for example, when $S = \emptyset$).

Sometime propositions take arguments, in which case they are called predicates.

> **Example 3.4**
>
> The predicate
> $$P(n) : n < n + 1$$
> is true for all natural numbers n.
> The predicate
> $$prime(n) : n \text{ is a prime number}$$
> is true for all primes $n = 2, 3, 5, 7, \ldots$ and false otherwise.

3.3 LOGICAL OPERATIONS

Similar to conditional statements in computer code, we use three operations to modify logical statements: and, or, and not.

Logical **conjunction** (and) is an operation that takes on the value true precisely when both the statements it modifies are true. It is denoted by the $\wedge$ symbol.

> **Example 3.5**
>
> The logical statement $(2 + 2 = 4) \wedge (1 + 1 = 2)$ is true since both $2 + 2 = 4$ and $1 + 1 = 2$ are true.
> The logical statement $(2 + 2 = 4) \wedge (1 + 2 = 4)$ is false since the second part of the statement is false.

Logical **disjunction** (or) is an operation that takes on the value true precisely when either of the statements it modifies are true. It is denoted by the $\vee$ symbol.

> **Example 3.6**
>
> The logical statement $(2 + 2 = 4) \vee (1 + 2 = 4)$ is true since the first part of the statement is true.
> The logical statement $(2 + 2 = 5) \vee (1 + 2 = 4)$ is false since both parts of the statement are false.

Logical **negation** (not), denoted by the ¬ symbol, is an operation on a single expression that flips true to false and false to true.

> **Example 3.7**
>
> The logical statement $\neg(2 + 2 = 5)$ is true as the statement $2 + 2 = 5$ is false. The logical statement $\neg(2 + 2 = 4)$ is false since $2 + 2 = 4$ is true.

We can use variables to serve as placeholders for logical statements.

> **Example 3.8**
>
> If we define P to be the logical statement $2 < 3$, then P is a true statement. It is also the case that $\neg\neg P$ is a true statement because negating a statement twice just gives back the original statement, i.e., $\neg\neg P$ is the same as P.

> **Example 3.9**
>
> For any logical statement P, $P \vee \neg P$ is true since it must be the case that either P or its complement is true. Similarly, $P \wedge \neg P$ is false since P or its complement is false.

In terms of order of operations, the negation operation has the highest precedence and all other ordering is shown using parentheses.

> **Example 3.10**
>
> In the expression $\neg A \wedge \neg B$ we first negate A and B and then perform the conjunction between the resulting values. On the other hand, for the expression $\neg(A \wedge B)$ the conjunction is performed first and then the result is negated. Note that it would never make sense to have an expression like $A \wedge B \vee C$ since it would be ambiguous whether to perform the conjunction or disjunction first.

3.4 TRUTH TABLES

A convenient way to represent expressions involving the above logical expressions is to use **truth tables**. The truth table for the operations from the previous section is given in Figure 3.1. The values true and false are abbreviated with T and F.

P	Q	$P \wedge Q$	$P \vee Q$	$\neg P$
F	F	F	F	T
F	T	F	T	T
T	F	F	T	F
T	T	T	T	F

Figure 3.1 Truth table for logical operations

You should create truth tables for more complex expressions by breaking them into smaller pieces.

Example 3.11

The truth table for $(A \wedge B) \vee \neg(B \wedge C)$ is broken down into smaller expressions and given in the table below. Notice that since there are now three variables (A, B, and C), we have $2^3 = 8$ rows in the table. Also notice that we shortlex order the rows so that we first set the leftmost variable to F and try all combinations of the remaining ones and only after that try all combinations with the leftmost variable set to T.

A	B	C	$A \wedge B$	$B \wedge C$	$\neg(B \wedge C)$	$(A \wedge B) \vee \neg(B \wedge C)$
F	F	F	F	F	T	T
F	F	T	F	F	T	T
F	T	F	F	F	T	T
F	T	T	F	T	F	F
T	F	F	F	F	T	T
T	F	T	F	F	T	T
T	T	F	T	F	T	T
T	T	T	T	T	F	T

The notation for and ($\wedge$) and or ($\vee$) may remind you of the notation for set intersection ($\cap$) and set union ($\cup$) and this is for good reason. When you compute the intersection of two sets, you are saying that the resulting elements must be in the first set **and** the second. Similarly, the elements of the union of two sets must be in the first set **or** the second.

The above connection means that some rules with sets generalize to logical statements. The following example shows that one of De Morgan's laws for sets applies for logic as well since the column for $\neg(A \vee B)$ is identical to the one for $\neg A \wedge \neg B$.

Example 3.12

The truth table below shows that $\neg(A \vee B) = \neg A \wedge \neg B$. Notice that to compute the column $\neg(A \vee B)$ we first had to create the column $A \vee B$; you should always break any expression down into smaller parts like this to show the intermediate steps.

A	B	$\neg A$	$\neg B$	$A \vee B$	$\neg(A \vee B)$	$\neg A \wedge \neg B$
F	F	T	T	F	T	T
F	T	T	F	T	F	F
T	F	F	T	T	F	F
T	T	F	F	T	F	F

Since $\neg(A \vee B)$ and $\neg A \wedge \neg B$ have identical columns in the truth table, we say that they are **logically equivalent**.

Definition 3.1

Two logical expressions are logically equivalent if they have identical truth values in the columns of a truth table.

As an exercise, show that $\neg(A \wedge B)$ and $\neg A \vee \neg B$ are logically equivalent (this is another of De Morgan's laws).

3.5   CONDITIONAL STATEMENTS

Another important logical operation is the conditional. This applies when one logical expression implies another and is denoted using the $\Rightarrow$ operator.

Example 3.13

If we let P be the statement "it is raining" and Q be the statement "it is cloudy," then we can write the statement "If it is raining, then it is cloudy" as $P \Rightarrow Q$.

The truth table for conditionals is given in Figure 3.2. For example, a true expression always implies a true expression, so $T \Rightarrow T$ is true and $T \Rightarrow F$ is false. However, a false expression can imply a false or true expression, so $F \Rightarrow T$ and $F \Rightarrow F$ are both true. Another way to say this is that the result of the operation is true if the first expression is false or if the second is true and this logical equivalence is demonstrated with $\neg A \vee B$ in the last column of the table.

A	B	$\neg A$	$\neg A \vee B$	$A \Rightarrow B$
F	F	T	T	T
F	T	T	T	T
T	F	F	F	F
T	T	F	T	T

Figure 3.2 Truth table for conditional operation

Some people are thrown off when they see that $F \Rightarrow T$ is true. An example helps to understand why this is the case.

Example 3.14

Suppose I promise my daughter "If you clean your room, I will give you $10." If P is the statement "my daughter cleaned her room" and Q is the statement "I give my daughter $10" then the promise can be written as $P \implies Q$. Now, if my daughter doesn't clean her room (P is false) and I still give her $10 ($Q$ is true), then she can certainly not accuse me of lying.

Example 3.15

Let A be the proposition that your code is correct and let B be the proposition that your code passes all your tests. Then, certainly, we believe that $A \Rightarrow B$, that is, correct code implies that it passes all your tests. However, it is possible for your code to be incorrect and still pass all your tests (perhaps because you missed some edge case), which is the case that $F \Rightarrow T$.

A common error made by novices to logic is to confuse a statement and its **converse**.

Definition 3.2

The converse of a statement $P \Rightarrow Q$ is $Q \Rightarrow P$.

A quick check of the truth tables for $P \Rightarrow Q$ and $Q \Rightarrow P$ will show you that these expressions are not logically equivalent. We can see this more intuitively from the following example.

Example 3.16

The statement "If it is raining, then it is cloudy" is a true statement since the rain must come from somewhere. However, it is not equivalent to "If it is cloudy, then it is raining" (a false statement) since we've all seen days when it is cloudy but not rainy.

A useful tool when proving conditionals later will be to use the **contrapositive** of a conditional statement.

Definition 3.3

The contrapositive of a statement $A \Rightarrow B$ is the statement $\neg B \Rightarrow \neg A$, which is logically equivalent.

The logical equivalence of $\neg B \Rightarrow \neg A$ and $A \Rightarrow B$ is demonstrated in Figure 3.3.

A	B	$\neg A$	$\neg B$	$\neg B \Rightarrow \neg A$	$A \Rightarrow B$
F	F	T	T	T	T
F	T	T	F	T	T
T	F	F	T	F	F
T	T	F	F	T	T

Figure 3.3 Logical equivalence of the contrapositive

Thus, the contrapositive of a true statement is always a true statement.

Let P be the statement "it is raining" and Q be the statement "it is cloudy." The statement $P \Rightarrow Q$ ("If it is raining, then it is cloudy") is true since the rain has to come from clouds. Hence the contrapositive $\neg Q \Rightarrow \neg P$ ("If it is not cloudy, it is not rainy") must also be true (there can't be rain without clouds).

One last logical operator that we will define is called the **biconditional**.

Definition 3.4

We say that $P \Leftrightarrow Q$ if both $P \Rightarrow Q$ and $Q \Rightarrow P$ are true.

Example 3.18

A number n is divisible by 6 if and only if it is divisible by 2 and 3. We can write this as follows:

$$(x \text{ is divisible by } 6) \Leftrightarrow ((x \text{ is divisible by } 2) \wedge (x \text{ is divisible by } 3)).$$

The truth table for the biconditional is given in Figure 3.4.

A	B	$A \Rightarrow B$	$B \Rightarrow A$	$(A \Rightarrow B) \wedge (B \Rightarrow A)$	$A \Leftrightarrow B$
F	F	T	T	T	T
F	T	T	F	F	F
T	F	F	T	F	F
T	T	T	T	T	T

Figure 3.4 Truth table for biconditional

3.6 QUANTIFIERS

While it is easy to use logical operators for finite sets, we need additional notation when making statements about larger sets. For this, we introduce two **quantifiers**, namely the universal or "for all" quantifier $\forall$ and the existential or "there exists" quantifier $\exists$.

Example 3.19

We can write the statement "The empty set is a subset of all sets" as $\forall$ sets $S, \emptyset \subseteq S$.

Example 3.20

We can write the statement "There exists a set with size zero" as $\exists$ set $S, |S| = 0$.

These quantifiers can be alternated to make more complex mathematical statements. It is important to note that the *order* of the quantifiers matters and changing the order can greatly change the meaning of the statement.

> **Example 3.21**
>
> The statement "For every natural number, there exists a bigger natural number" can be written $\forall x \in \mathbb{N}, \exists y \in \mathbb{N}, y > x$.
>
> Note that the above statement is very different from $\exists y \in \mathbb{N}, \forall x \in \mathbb{N}, y > x$ which says that "There exists a natural number that is bigger than all natural numbers," which is clearly not true.

3.7 BIG-O NOTATION

Quantifiers are the basis of asymptotic notation that is used ubiquitously in computer science. You may have encountered Big-O notation when analyzing the running time of an algorithm you were studying. You'll get a taste of how this is defined using quantifiers in this section.

Say that you have an algorithm that on input of size n performs $f(n) = 2n + 3$ computational operations. For practical purposes, the $+3$ in the formula is not very relevant as it is going to contribute very little when n gets large. Even the coefficient of 2 in the formula is not very important when we want to study the growth of the function f. In Big-O notation, we say that f is $O(n)$ or linear in n.

> **Example 3.22**
>
> The function $g(n) = 3n^2 + 5n + 3$ is $O(n^2)$.

The mathematical definition of Big-O is as follows:

> **Definition 3.5**
>
> We say that $f(n)$ is $O(g(n))$ if there exists $n_0, c > 0$ such that for all $n \geq n_0$ we have that $f(n) \leq cg(n)$. This can be written as
>
> $$\exists n_0, c > 0, \forall n \geq n_0, f(n) \leq cg(n).$$

In other words, we can say that the function f is bounded by g times some constant factor for sufficiently large values of n. This definition allows us to categorize the growth rate of functions into ones such as linear ($O(n)$) or quadratic ($O(n^2)$).

> **Example 3.23**
>
> The function $f(n) = 2n+4$ is $O(n)$ using $c = 3$ and $n_0 = 4$, that is, $2n+4 \leq 3n$ for all $n \geq 4$. This is true because $4 \leq n$ for all $n \geq 4$, so $2n+4 \leq 2n+n = 3n$ for all $n \geq 4$. Note that any larger value of c and n_0 would also work.

Example 3.24

> The function $f(n) = n^2 + 4n + 2$ is $O(n^2)$ using $c = 7$ and $n_0 = 1$, that is, $n^2 + 4n + 2 \leq 7n^2$ for all $n \geq 1$. This is true because $4n \leq 4n^2$ for all $n \geq 1$ and $2 \leq 2n^2$ for all $n \geq 1$, so $n^2 + 4n + 2 \leq n^2 + 4n^2 + 2n^2 = 7n^2$ for all $n \geq 1$.

For more on asymptotic notation, see Appendix D.

3.8 NEGATING LOGICAL STATEMENTS

You will at times need to negate logical statements. For example, negating statements is really important for writing `while` loops because you will want to negate the stopping condition to get the `while` condition.

To negate unquantified statements, we can apply De Morgan's laws to get that the negation of $P \wedge Q$ is $\neg P \vee \neg Q$ and the negation of $P \vee Q$ is $\neg P \wedge \neg Q$

Example 3.25

> The negation of $(x \geq 2) \wedge (x \leq 5)$ is $\neg(x \geq 2) \vee \neg(x \leq 5)$, which can be re-written as $(x < 2) \vee (x > 5)$. Note that the negation of $\geq$ is $<$ and the negation of $\leq$ is $>$.
>
> This example shows us how to create a `while` loop condition if we want to input a number between 2 and 5 inclusive. The loop condition should be to ask the user to input numbers x while x < 2 or x > 5.

Example 3.26

> The negation of $(x = 1) \vee (x = 2)$ is $\neg(x = 1) \wedge \neg(x = 2)$, which can be re-written as $(x \neq 1) \wedge (x \neq 2)$.

To negate quantified statements, we have to flip universal ($\forall$) and existential ($\exists$) quantifiers before negating the innermost statement.

Example 3.27

> The negation of the true statement "For every natural number, there exists a bigger natural number"
>
> $$\forall x \in \mathbb{N}, \exists y \in \mathbb{N}, y > x$$
>
> is
>
> $$\exists x \in \mathbb{N}, \forall y \in \mathbb{N}, y \leq x,$$
>
> which says "There exists a natural number that is bigger than or equal to all natural numbers," which is a false statement.

Example 3.28

The negation of

$$\exists n_0, c \in \mathbb{N}, \forall n \geq n_0, f(n) \leq cg(n)$$

is

$$\forall n_0, c \in \mathbb{N}, \exists n \geq n_0, f(n) > cg(n).$$

3.9   CHAPTER SUMMARY AND KEY CONCEPTS

- A **logical statement** is a mathematical statement that takes on the values **true** or **false**.

- Commonly used **logical operations** performed on logical statements include **conjunction (and)**, **disjunction (or)**, and **negation (not)**.

- **Truth tables** can be used to compute logical operations and to show that two logical expressions are **logically equivalent**.

- **Conditional statements** are used for logical conditionals or implication. The **contrapositive** of a statement is logically equivalent to the original statement. A conditional statement is different from its **converse**.

- The **for all** ∀ **quantifier** and **there exist** ∃ **quantifier** allow us to make quantified statements for the elements of a set.

- **Big-O notation** allows us to measure the rate of growth of functions, with applications such as expressing the running time of an algorithm.

- Logical statements can be **negated** by carefully switching ∀/∃ and negating the innermost statement.

EXERCISES

Exercises for 3.2

Determine whether each of the following statements is true or false. Give a one line explanation for each.

3.1 (Assume that $x = 3$) $x \leq 2$

3.2 (Assume that $x = 3$) $x < 3$

3.3 (Assume that $x = 3$) $(x \leq 3) \wedge (x \leq 2)$

3.4 (Assume that $x = 3$) $(x \leq 3) \vee (x \leq 2)$

3.5 (Assume that $x = 3$ and $y = 4$) $(x < y) \wedge (x + 1 \leq y)$

3.6 Every number is either even or odd.

3.7 For every number, there is always a bigger number.

3.8 For every natural number, there is always a smaller natural number.

3.9 For all integers x and y, $x + y$ is an integer.

3.10 For all integers x and y, x / y is a real number.

Exercises for 3.3

Write each of the following statements in symbolic form in terms of $\neg$, $\wedge$, and $\vee$ operations.

3.11 The value x is bigger than zero but smaller than ten.

3.12 The value x is at least zero and at most ten.

3.13 The value x is smaller than zero or at least eleven.

3.14 The value x is not bigger than six.

3.15 The value x is not bigger than six and is not smaller than four.

3.16 All three values x, y, z are equal to ten.

3.17 At least one of the values x, y, z is equal to ten.

3.18 Exactly one of the values x, y, z is equal to ten.

3.19 Exactly two of the values x, y, z are equal to ten.

3.20 The value $year$ is divisible by four, but when it is divisible by 100 it must also be divisible by 400. (This is the condition for a year to be a leap year.) You can use the % operation for computing the remainder from division.

3.21 The value $year$ is not a leap year. (See the previous problem for computing if a year is a leap year.)

Exercises for 3.4

Show whether the following pairs of expressions are equivalent by creating truth tables for them. Be sure to show each of the intermediate steps for each expression.

3.22 $A \wedge A$ and A

3.23 $A \vee A$ and A

3.24 $A \wedge B$ and $B \wedge A$

3.25 $A \vee B$ and $B \vee A$

3.26 $A \wedge (A \vee B)$ and B

3.27 $A \wedge (A \vee B)$ and A

3.28 $A \vee (A \wedge B)$ and B

3.29 $A \vee (A \wedge B)$ and A

3.30 $(A \wedge B) \wedge C$ and $A \wedge (B \wedge C)$

3.31 $(A \vee B) \vee C$ and $A \vee (B \vee C)$

3.32 $A \wedge (B \vee C)$ and $(A \wedge B) \vee (A \wedge C)$

3.33 $A \vee (B \wedge C)$ and $(A \vee B) \wedge (A \vee C)$

Exercises for 3.5

Write each of the following statements in the form $P \Rightarrow Q$ or $P \Leftrightarrow Q$, stating the values of P and Q for each.

3.34 Your code works correctly precisely when it passes all your test cases.

3.35 If a number is divisible by 4 it must be divisible by 2.

3.36 The sum of two numbers is an integer if both numbers are integers.

3.37 It is safe to divide by x if $x \neq 0$.

3.38 It is safe to divide by x only if $x \neq 0$.

Show whether the following pairs of expressions are equivalent by creating truth tables for them. Be sure to show each of the intermediate steps for each expression.

3.39 $A \vee B$ and $\neg B \Rightarrow A$

3.40 $A \vee B$ and $\neg A \Rightarrow B$

3.41 $A \wedge B$ and $A \Rightarrow \neg B$

3.42 $A \Leftrightarrow B$ and $(\neg A \vee B) \wedge (A \vee \neg B)$

3.43 $(A \Rightarrow B) \wedge (A \Rightarrow C)$ and $A \Rightarrow (B \wedge C)$

3.44 $(A \Rightarrow B) \vee (A \Rightarrow C)$ and $A \Rightarrow (B \vee C)$

3.45 $(A \Rightarrow C) \wedge (B \Rightarrow C)$ and $(A \wedge B) \Rightarrow C$

3.46 $(A \Rightarrow C) \wedge (B \Rightarrow C)$ and $(A \vee B) \Rightarrow C$

Exercises for 3.6

Write the following as English statements with no mathematical notation.

3.47 $\forall n \in \mathbb{N}, 2n \in \mathbb{N}$

3.48 $\forall n \in \mathbb{R}, n/2 \in \mathbb{R}$

3.49 $\forall S \in \mathcal{P}(A), S \subseteq A$

3.50 $\forall n \in \mathbb{Z}, \exists m \in Z, m < n$

3.51 $\forall n \in \mathbb{N}, \exists l, m \in \mathbb{N}, n = l + m$

3.52 $\forall n \in \mathbb{N}, \exists s \in \{0, 1\}^*, |s| = n$

3.53 $\forall s \in \{0, 1\}^*, \exists q, r \in \{0, 1\}^*, s = qr$

Exercises for 3.7

Show the following Big-O relationships by giving constants c and n_0 in each case and showing that they work.

3.54 $3n + 4 \in O(n)$

3.55 $7n - 2 \in O(n)$

3.56 $2n^2 + 4 \in O(n^2)$

3.57 $3n^2 - 2n + 4 \in O(n^2)$

3.58 $n^3 - n^2 + n - 1 \in O(n^3)$

Exercises for 3.8

Rewrite the following statements into mathematical form and then negate them.

3.59 The value x is bigger than zero and smaller than ten.

3.60 The values x and y are both equal to ten.

3.61 The values x, y, z are all equal.

3.62 The values x, y, z are all different.

3.63 The Boolean value $hasWon$ or $hasLost$ is true

3.64 The Boolean value $hasWon$, $hasLost$, or $isTied$ is true

Negate the following quantified statements.

3.65 $\forall n \in \mathbb{R}, 2n \in \mathbb{R}$

3.66 $\exists n \in \mathbb{Z}, 1/n \notin \mathbb{R}$

3.67 $\forall n \in \mathbb{N}, \exists m \in \mathbb{N}, m < n$

3.68 $\forall n \in \mathbb{N}, \exists l, m \in \mathbb{N}, n = l + m$

3.69 $\forall n \in \mathbb{N}, \exists s \in \{0,1\}^*, |s| = n$

3.70 $\forall s \in \{0,1\}^*, \exists q, r \in \{0,1\}^*, s = qr$

3.71 $\exists w \in \Sigma^*, \forall x, y \in \Sigma^*, \exists z \in \Sigma^*, (|wx| \leq |yz|) \vee (|zx| \leq |yw|)$

CHAPTER 4

Nondeterministic Finite Automata

4.1 WHY YOU SHOULD CARE

Nondeterministic Finite Automata (NFAs) allow us to create simpler, often smaller, automata for the same language. They are a generalization of DFAs that add in a property called nondeterminism. While nondeterminism is not something that any real computer can exhibit, it is an important concept for much of theoretical computer science. In particular, in later courses you will learn about hard NP-complete problems for which nondeterminism is an integral concept to understand.

4.2 WHY NFAS CAN BE SIMPLER THAN DFAS

We will start with an example of a language that is tricky to get a correct DFA for and then see how to create an exceedingly simple NFA for it. Consider the language $L = \{s \in \{0,1\}^* : s \text{ ends in } 010101\}$. A DFA for this language is given below:

Example 4.1

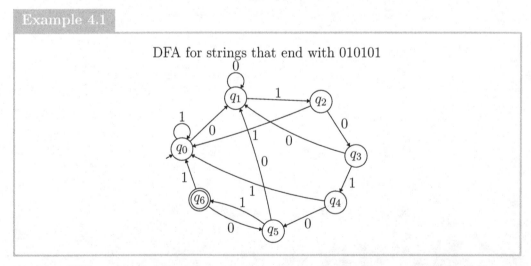

DFA for strings that end with 010101

This DFA works by having states for strings that end in each of the following prefixes of 010101: 0, 01, 010, 0101, 01010, and 010101 (states q_1 through q_6). However,

DOI: 10.1201/9781003383284-4

you have to be careful about where each transition goes to maintain the correct prefix property for each state, making the diagram fairly complicated.

In contrast, consider an NFA for the same language in the following example:

Example 4.2

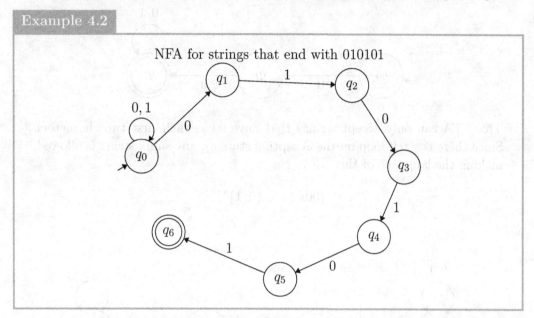

NFA for strings that end with 010101

Let us use this example to understand a few differences between NFAs and DFAs. You may first notice that the state q_0 has two possible destinations on a 0. Nondeterminism means that the NFA is able to follow either transition. The diagram is also a lot simpler because there are no transitions for many state/symbol pairs—for example, there is no transition from q_1 on a 0. These are two of the major differences between NFAs and DFAs. The other big difference to understand is how strings are accepted.

An NFA accepts a string if there exists a legal path from q_0 to an accepting state while traversing the string. For example, the string 010101 is accepted by the NFA since it can follow the transitions to q_1 on the first 0, to q_2 on the first 1, and so on, until it reaches the accepting state q_6. Importantly, there may be other paths that don't lead to an accepting state, but this doesn't change the fact that an accepting path exists. For example, an alternate processing of string 010101 would be to stay at q_0 and thus reject. Yet another rejecting path would be to stay at q_0 for 0101 and then get to q_2 on the final 01. However, since there does exist at least one accepting path, we ignore all the rejecting paths and conclude that 010101 is accepted by the NFA.

Now, to understand why this NFA accepts the correct language, we will see that it accepts all strings that end in 010101 and no others. For any string that does end in 010101, we can self-loop at q_0 until such time that we arrive at the sixth last symbol and then follow the transitions to q_6. Conversely, the only strings that can be accepted must end in 010101 as that is the sequence that must be followed to arrive at the only accepting state q_6. Also note that there are no transitions out of q_6, so the string really must end in 010101 to be accepted.

4.3   MORE EXAMPLE NFAS

Example 4.3

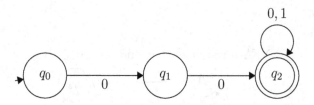

This NFA can only accept strings that have 00 as their first two characters. Since there is a self-loop on the accepting state q_2, any suffix string is allowed, making the language of this NFA

$$\{00s : s \in \{0,1\}^*\}.$$

Example 4.4

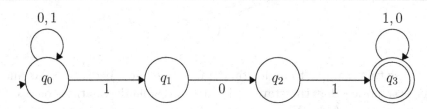

This NFA can have any prefix and any suffix over $\{0,1\}^*$ but it must contain the substring 101 somewhere in it, giving the language:

$$\{p101s : p,s \in \{0,1\}^*\}.$$

Example 4.5

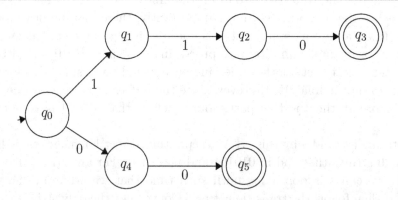

This NFA accepts the language $\{110, 00\}$ with only two strings.

This last example shows us a way to easily create an NFA for any finite language.

Example 4.6

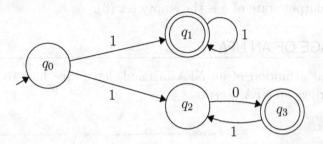

This NFA rejects any string that doesn't start with a 1 since the only transitions out of the start state are on a 1. On the upper part of the NFA it goes to a state that accepts strings with one or more 1s in it. The lower part accepts precisely the strings that alternate 10s one or more times. Thus the language of the above NFA is $\{1^n : n \geq 1\} \cup \{(10)^n : n \geq 1\}$.

This example shows an easy way to design an NFA that accepts the union of two regular languages—we can simply create NFAs for the two languages and then create an NFA that nondeterministically chooses between them so that the strings from both languages are accepted.

4.4 FORMAL DEFINITION OF AN NFA

Now that you have seen some examples, we can see a formal definition for an NFA. An NFA is formally defined by a 5-tuple $(Q, \Sigma, \delta, q_0, F)$, where Q is the set of states, Σ is the input alphabet, $\delta : Q \times \Sigma \to \mathcal{P}(Q)$ is the transition function, q_0 is the start state, and F is the set of final or accepting states.

Example 4.7

The NFA in Example 4.6 is given by the tuple $(Q, \Sigma, \delta, q_0, F)$, where

- $Q = \{q_0, q_1, q_2, q_3\}$

- $\Sigma = \{0, 1\}$

- δ is given by the table:

	0	1
q_0	$\emptyset$	$\{q_1, q_2\}$
q_1	$\emptyset$	$\{q_1\}$
q_2	$\{q_3\}$	$\emptyset$
q_3	$\emptyset$	$\{q_2\}$

- $F = \{q_1, q_3\}$

Note that the mathematical definition is nearly identical to that of a DFA and only δ is really different. The δ function's output is now a *set* of states since transitions can now go to multiple states from any state/symbol pair. In the case that a transition is missing, the output state of δ is the empty set ($\emptyset$).

4.5   LANGUAGE OF AN NFA

With the formal definition of an NFA in hand, let us see how to mathematically define which strings an NFA accepts.

Definition 4.1

Let some string $s \in \Sigma^*$ of length $n \geq 0$ be written as $s = s_1 s_2 s_3 \ldots s_n$, where each $s_i \in \Sigma$ for $1 \leq i \leq n$. We say that an NFA $(Q, \Sigma, \delta, q_0, F)$ accepts the string s precisely when

$$\exists p_0 = q_0, p_1, p_2, \ldots p_n \in Q, (\forall i \in \{0, \ldots, n-1\}, p_{i+1} \in \delta(p_i, s_i)) \wedge (p_n \in F).$$

At first glance the above quantified formula looks quite intimidating. However, we can break it down to see that the idea is not very complex. It says that the string $s = s_1 s_2 s_3 \ldots s_n$ will be accepted if we can find a sequence of states p_0 through p_n (where p_0 is just defined to be the start state q_0) such that after each symbol s_i ($1 \leq i \leq n$) we can use a transition in δ that takes us from state p_{i-1} to state p_i on symbol s_i and that the last state we reach (p_n) must be an accepting state. We can see this using an example.

Example 4.8

Consider the following NFA that accepts the language

$$\{xaby : x, y \in \{a, b\}^*\}.$$

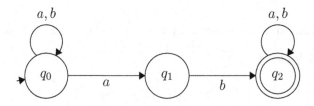

We can see that this NFA accepts the string $aaaba$ because of the existence of a sequence of states $q_0, q_0, q_0, q_1, q_2, q_2$. We can verify that starting at state q_0, we can reach each of the states q_0, q_0, q_1, q_2, q_2 on the symbols a, a, a, b, a and that the last state (q_2) is accepting. On the other hand, the NFA will not accept the string $bbaa$ since it is impossible to find a sequence of states that allow us to go from q_0 to the accepting state q_2 using only transitions defined for the NFA.

Example 4.9

Consider the NFA in Example 4.6. We can see that it accepts the string 1010 because of the existence of the sequence of states q_0, q_2, q_3, q_2, q_3. We can see that starting at state q_0 we can reach the states q_2, q_3, q_2, and q_3 on symbols 1, 0, 1, and 0, and that q_3 is accepting. Similarly, the string 111 is accepted because of the sequence of states q_0, q_1, q_1, q_1. On the other hand, the NFA cannot accept 1011 because there is no sequence of states that allow us to go from q_0 to one of the accepting states (q_1 or q_3) on the symbols 1, 0, 1, 1.

4.6 SUBSET CONSTRUCTION

The NFA is what is called a generalization of the DFA because it can do all that a DFA can do. In particular, we can think of a DFA as an NFA that has exactly one transition defined for each state/symbol pair. This might seem to imply that NFAs are more powerful than DFAs. However, we will see next that DFAs are capable of accepting all the languages that NFAs can accept showing that they are in fact equivalent in the languages that they can accept.

We will show that DFAs are as powerful as NFAs constructively by giving an algorithm that can convert any NFA into a DFA with the same language. This algorithm is known as the Subset Construction algorithm. For any NFA, we construct a DFA that has as its states sets of states from the NFA. The complete Subset Construction procedure is given in Algorithm 4.1.

Algorithm 4.1: SubsetConstruction(N)

input : An NFA $N = (Q_N, \Sigma, \delta_N, q_0, F_N)$
output: A DFA that accepts precisely the same language as N
1: Set $Q_D = \emptyset$
2: Create the state $\{q_0\}$ and add it to Q_D
3: **while** there is some $q \in Q_D$ and $s \in \Sigma$ without a transition defined **do**
4:    Let q' be the union of all the states reachable in Q_N from each of the states in q on symbol s
5:    **if** $q' \notin Q_D$ **then**
6:        Add q' to Q_D
7:    Add the transition $\delta_D(q, s) = q'$
8: Set $F_D = \emptyset$
9: **for all** $q \in Q_D$ **do**
10:    **if** any state in q is in F_N **then**
11:        Add q to F_D
12: Return the constructed DFA $D = (Q_D, \Sigma, \delta_D, \{q_0\}, F_D)$

The algorithm works by creating a DFA that has as states the *sets of states* of the NFA. Intuitively, each state of the DFA (a set of states of the NFA) represents the set of states that a string can reach. We then have to add in the transitions from each state appropriately. If the number of states of the DFA is n, we may have as many as 2^n subsets of these states. However, in practice the number we need often turns out to be a lot smaller. We only create the states that are necessary by adding in new states as needed. We start at the state $\{q_0\}$ since q_0 is the unique starting state of the NFA and add any new states reachable from it. We continue this process until there are no new states and every state has all its transitions defined for it.

Perhaps the easiest way to convert an NFA to a DFA is by using the δ transition table. The table for the NFA in Example 4.3 is:

	0	1
q_0	$\{q_1\}$	$\emptyset$
q_1	$\{q_2\}$	$\emptyset$
q_2	$\{q_2\}$	$\{q_2\}$

The table for the DFA from subset construction is given below. We start with the state $\{q_0\}$ since q_0 is the unique starting state. On a 0 from q_1 we can only go to q_1, so we add a transition to the state $\{q_1\}$. On a 1 from q_0 we can go nowhere, so we add a transition to the *state* $\emptyset$. The empty set state is special in that it contains none of the states of the NFA, so the only place we can go from it is to itself (i.e., it is a sink state). The transitions from $\{q_1\}$ and $\{q_2\}$ are the same as for q_1 and q_2. We thus get the δ transition table below:

	0	1
$\{q_0\}$	$\{q_1\}$	$\emptyset$
$\emptyset$	$\emptyset$	$\emptyset$
$\{q_1\}$	$\{q_2\}$	$\emptyset$
$\{q_2\}$	$\{q_2\}$	$\{q_2\}$

Note that the state $\{q_2\}$ will be the only accepting state as it is the only one that contains an accepting state of the NFA (namely, q_2). The final diagram looks as follows:

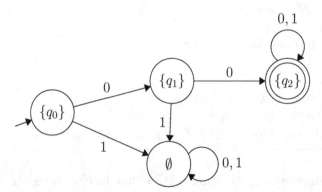

Example 4.11

With some practice, you can also convert an NFA from the diagram. We will convert the NFA in Example 4.2 into a DFA. We start with the state $\{q_0\}$ and see that on a 0 we can go to the states $\{q_0, q_1\}$ so we add the new state $\{q_0, q_1\}$ to our DFA with a transition to it from $\{q_0\}$ on a 0. On a 1 from $\{q_0\}$ we can only go back to $\{q_0\}$, so we add a self-loop of 1 on $\{q_0\}$.

Next, starting from $\{q_0, q_1\}$ we can only go to the states $\{q_0, q_1\}$ on a 0 since q_0 still goes to $\{q_0, q_1\}$ on a 0 and q_1 goes nowhere ($\emptyset$) and the union of these is $\{q_0, q_1\}$. On a 1 from $\{q_0, q_1\}$ we can go to $\{q_0, q_2\}$ since we can go to $\{q_0\}$ from q_0 and to $\{q_2\}$ from q_1.

From $\{q_0, q_2\}$ on a 0 we can get to $\{q_0, q_1, q_3\}$ ($\{q_0, q_1\}$ from q_0 and $\{q_3\}$ from q_2). From $\{q_0, q_2\}$ on a 1 we can only get back to $\{q_0\}$ since q_0 goes to $\{q_0\}$ and q_2 goes nowhere on a 1.

Continuing this procedure, we get the DFA below. The only accepting state is $\{q_0, q_2, q_4, q_6\}$ as it is the only one that contains the accepting state from the NFA (q_6). Notice that it is identical to the original DFA we had in Example 4.1 with only the state names different.

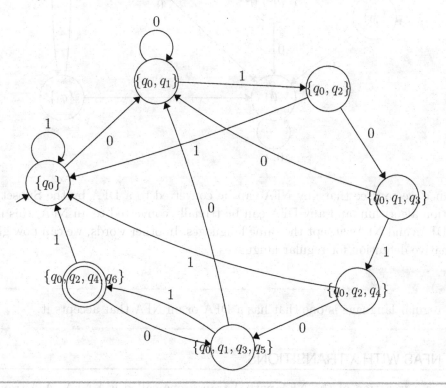

Example 4.12

Consider the NFA from Example 4.6. We start with the starting state $\{q_0\}$ and see that on a 0 we cannot go to any states, so we create a new state $\emptyset$ and add a transition from $\{q_0\}$ to $\emptyset$. We can reach both q_1 and q_2 from $\{q_0\}$ on a 1, so we add the new state $\{q_1, q_2\}$ and transition to it from $\{q_0\}$ on a 1. Since we cannot get to any states from the empty set, we add a self-loop to it on both 0 and 1. Continuing this procedure, we get the DFA below. The accepting states are $\{q_1, q_2\}$, $\{q_3\}$, and $\{q_1\}$ since these are the only states that contain q_1 or q_3, the accepting states of the original NFA.

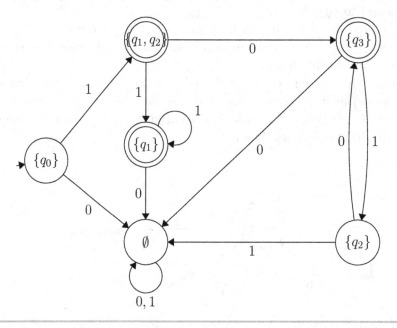

Since we can see that any NFA can be converted to a DFA by the Subset Construction algorithm, and any DFA can be trivially converted to an NFA, this means that DFAs and NFAs accept the same languages. In other words, we can now give an alternative definition for regular languages.

Definition 4.2

A regular language is one that has a DFA or an NFA that accepts it.

4.7 NFAS WITH λ TRANSITIONS

There is a variation of the NFA with λ-transitions called a λ-NFA. A λ-transition can be followed while not consuming a character of the string. For example, if the language is the union of two languages with simple NFAs, we can create an NFA with the union of their languages by adding initial λ-transitions, as seen in the following example:

Example 4.13

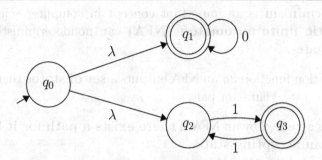

Any string accepted by this λ-NFA can first nondeterministically transition to either q_1 or q_2 without consuming any characters. From q_1 only the strings in $\{0\}^*$ are accepted. From q_2 we can accept only the strings with an odd number of 1s. Thus the language for this NFA is the union of these languages:

$$\{0\}^* \cup \{1^{2n+1} : n \in \mathbb{N}\}.$$

The formal definition of an λ-NFA is identical to that of an NFA except that we now add transition rules for δ on each state on a λ as seen in this example.

Example 4.14

The NFA in Example 4.13 is given by the tuple $(Q, \Sigma, \delta, q_0, F)$, where

- $Q = \{q_0, q_1, q_2, q_3\}$

- $\Sigma = \{0, 1\}$

- δ is given by the table:

	0	1	λ
q_0	$\emptyset$	$\emptyset$	$\{q_1, q_2\}$
q_1	$\{q_1\}$	$\emptyset$	$\emptyset$
q_2	$\emptyset$	$\{q_3\}$	$\emptyset$
q_3	$\emptyset$	$\{q_2\}$	$\emptyset$

- $q_0 = q_0$

- $F = \{q_1, q_3\}$

We can also convert a λ-NFA into a DFA using a procedure similar to the Subset Construction algorithm. The only difference will be that we have to make sure to also follow λ-transitions when determining what set of states can be reached from any state.

4.8   CHAPTER SUMMARY AND KEY CONCEPTS

- **Nondeterminism** is an important concept in computer science. A **nondeterministic finite automaton (NFA)** can nondeterministically transition between states.

- The transition function for an NFA outputs a **set of states** that can be reached on a given state, character pair.

- A string is accepted by an NFA if **there exists a path for it from the start state to an accepting state**.

- The **Subset Construction** algorithm can be used to convert any NFA into a DFA.

- A **regular language** is one that has a DFA or an NFA that accepts it.

- A variation of the NFA, called the λ-**NFA**, allows for λ-transitions to go from a state to another state without consuming a character of the string.

EXERCISES

Give a sequence of states that show that the following strings are accepted by the following NFAs:

4.1 10010101 for the NFA in Example 4.2

4.2 0011 for the NFA in Example 4.3

4.3 01010 for the NFA in Example 4.4

4.4 1111 for the NFA in Example 4.6

4.5 101010 for the NFA in Example 4.6

4.6 ababab for the NFA in Example 4.8

4.7 baaab for the NFA in Example 4.8

Draw nondeterministic finite automata (that are not also deterministic ones) with no λ transitions for the following languages:

4.8 $\{w \in \{0,1\}^* : w \text{ ends in } 00\}$

4.9 $\{w \in \{0,1\}^* : w \text{ has } 01 \text{ as a substring}\}$

4.10 $\{w \in \{0,1\}^* : w \text{ has } 01 \text{ or } 10 \text{ as a substring}\}$

4.11 $\{w \in \{a,b\}^* : w \text{ ends in ab or ba}\}$

4.12 $\{w \in \{0,1\}^* : w \text{ starts with a } 0 \text{ and ends with a } 1\}$

4.13 $\{w \in \{0,1\}^* : w \text{ starts with } 00 \text{ and ends with } 11\}$

4.14 $\{a\}^* \cup \{(ab)^n : n \geq 1\}$

4.15 $\{a^n b^m : n \geq 0, m \geq 1\}$

4.16 $\{0^p 1^q 0^r : p \geq 1, q \geq 0, r \geq 1\}$

4.17 $\{w \in \{0,1\}^* :$ the third character from the end in w is $0\}$

4.18 $\{w \in \{0,1\}^* :$ at least one of the last three characters from the end is a $0\}$

4.19 $\{w \in \{0,1\}^* : w$ does not have 01 as a substring$\}$

Convert the following NFAs to DFAs using the Subset Construction algorithm:

4.20 The NFA in Example 4.4

4.21 The NFA in Example 4.5

4.22 The NFA in Example 4.8

CHAPTER 5

Regular Expressions

5.1 WHY YOU SHOULD CARE

Regular expressions are a shorthand way to express regular languages. By using combinations of just three operations, we can express any regular language. In addition to their connection with regular languages, regular expressions are used extensively in string matching in text search and modern text editors, as well as for search utilities for the Unix command line.

5.2 WHY REGULAR EXPRESSIONS

Suppose that you are doing a research project in which you are looking up all winners of the Turing award (the computer science equivalent of the Nobel Prize) and you find a web page that has the list on it. You'd like to get your hands on the list, but it is encoded using HTML (HyperText Markup Language) that looks something like this:

```
<li><span>1966</span><a href="/perlis">Perlis, Alan J</a></li>
<li><span>1967</span><a href="/wilkes">Wilkes, Maurice V</a></li>
<li><span>1968</span><a href="/hamming">Hamming, Richard W</a></li>
<li><span>1968</span><a href="/minsky">Minsky, Marvin L</a></li>
...
```

You figure out the code for scraping this web page using your language of choice and now have the task of extracting the names from this HTML. While it would be a good problem-solving exercise with string processing to get each of the names, a far better way to solve this problem is to use a tool called regular expressions. By simply making the observation that each line above is of the form `<li><span>YEAR</span><a href="URL">NAME</a></li>`, where `YEAR` is the year they won the award, `URL` is the URL (uniform resource locator) for their page, and `NAME` is the name of the winner, you can get all the names easily with a couple lines of code. You'll see how to do this later in this chapter.

DOI: 10.1201/9781003383284-5

5.3 REGULAR EXPRESSION OPERATIONS

The three operations for regular expressions are union (+), concatenation (·), and Kleene star (*). The union operation is identical to set union.

Example 5.1

The regular expression $01 + 10$ is shorthand for the language $\{01, 10\}$. The regular expression $0 + 1 + 00$ represents the language $\{0, 1, 00\}$.

The concatenation operation allows for concatenation of two regular expressions to represent the set of strings that have a string from the first expression followed by a string from the second.

Example 5.2

The regular expression $(0 + 1) \cdot 1$ represents the language $\{01, 11\}$. We will usually omit the · symbol as in the next example. The regular expression $(0 + 1)(0 + 11)$ represents the language $\{00, 011, 10, 111\}$.

Finally, the Kleene star operation on a regular expression generates the set of strings consisting of zero or more strings from the regular expression.

Example 5.3

The expression 0^* represents the set $\{\lambda, 0, 00, 000, \ldots\}$. The expression $(0+10)^*$ represents the set

$$\{\lambda, 0, 00, 10, 000, 010, 100, 0000, 0010, 0100, 1000, 1010, \ldots\}.$$

It is worth paying attention to the precedence of the above three operations. The Kleene star operation has the highest precedence in that it applies to the smallest sequence of characters before it that make up a valid regular expression. The concatenation operation is next in precedence and union comes last. Parentheses can always be added to group expressions.

Example 5.4

In the expression $10(0 + 1)^* + 01$, the Kleene * applies to the smallest valid expression to the left of it, $(0 + 1)$ in this example (as opposed to $10(0 + 1)$, say). The concatenation of 10 to $(0 + 1)^*$ takes higher precedence than the union with 01. Thus, the expression $10(0 + 1)^* + 01$ represents the set $\{01, 10, 100, 101, 1000, 1001, \ldots\}$.

You may have noticed that in all the examples above the expression is said to represent the set. This is an important distinction—the type of a regular expression is not a set, but its own type that represents a set. More specifically, we will use the notation $L(r)$ to represent the language of a regular expression r.

Example 5.5

The following are some examples of languages of regular expressions:

1. $L((0+1)^*) =$ all strings over $\{0,1\}$

2. $L((a+b)^*a(a+b)^*) =$ all strings over $\{a,b\}$ with at least one a

3. $L(((a+b)(a+b))^*) =$ all strings over $\{a,b\}$ of even length

For convenience, we allow shorthand notation for strings to appear in regular expressions. Note that the exponent must always be a fixed natural number—we do not allow variables in the exponent.

Example 5.6

The expression $a^3(a+b)$ is identical to $aaa(a+b)$. The expression $(a+b)^4$ is identical to $(a+b)(a+b)(a+b)(a+b)$ and represents the set of all strings of length exactly 4 over $\{a,b\}$.

5.4 FORMAL DEFINITION OF REGULAR EXPRESSIONS

Regular expressions are defined recursively as follows.

Definition 5.1

1. The expression $\emptyset$ is a regular expression that represents the set $\emptyset$.

2. The expression λ is a regular expression that represents the set $\{\lambda\}$.

3. For each symbol $s \in \Sigma$, the expression s is a regular expression that represents the set $\{s\}$.

4. For any regular expressions r_1 and r_2, the expression $r_1 + r_2$ is a regular expression that represents the set $L(r_1) \cup L(r_2)$.

5. For any regular expressions r_1 and r_2, the expression $r_1 \cdot r_2$ is a regular expression that represents the set $\{s_1 \cdot s_2 : s_1 \in L(r_1) \wedge s_2 \in L(r_2)\}$.

6. For any regular expression r_1, the expression $(r_1)^*$ is a regular expression that represents the set $\{s_1 s_2 s_3 \ldots s_n : n \geq 0 \wedge s_1, s_2, s_3, \ldots, s_n \in L(r_1)\}$.

We can build fairly complex expressions using these rules, as seen next.

Example 5.7

- The empty set is a regular language as it has the regular expression $\emptyset$.

- The set $\{\lambda\}$ is a regular language as it has the regular expression λ.

- The set $\{a\}$ is a regular language as it has the regular expression a.

- The set $\{a, b\}$ is a regular language as it has the regular expression $a + b$.

- The set $\{aa, ab, ba, bb\}$ is a regular language as it has the regular expression $(a + b)(a + b)$.

- The set $\{\lambda, aa, aaaa, aaaaaa, \ldots\}$ is a regular language as it has the regular expression $(aa)^*$.

- The expression $(aa)^*(a + b)(a + b) + \lambda$, which represents the set $\{\lambda, aa, ab, ba, bb, aaaa, aaab, aaba, aabb, \ldots\}$, is made up of smaller expressions: $a, aa, (aa)^*, b, (a + b), (aa)^*(a + b), (aa)^*(a + b)(a + b)$, and λ.

5.5 APPLICATIONS

Many programming languages have support for regular expressions as you can use them to search through large bodies of text to find substrings that are formatted like phone numbers or addresses. Several advanced text editors will also allow you to perform searches using regular expressions.

Regular expressions are invaluable when you are learning the command line in Linux. For example, say that you want to search for all files in a folder that start with the letter a. You would do this typing `ls a*` into the terminal. This invokes Linux's list command `ls` and specifies that only the files that have an `a` at the start should be shown. Here, the * symbol corresponds to the regular expression Σ^*, so any string (including the empty string) can match the end of the filename. Similarly, if you want all files that start with an `a` and end in the extension `.txt`, you would use the following command: `ls a*.txt`. If you want to find all files that contain the string `Lall`, you would use this command: `ls *Lall*`.

You also have the ability to match single characters using the symbol ?. For example, to match all filenames that start FILE and then have a single character, you could use: `ls FILE?` or if you want all files that have names exactly three characters long, you would use `ls ???`.

5.6 REGULAR EXPRESSIONS IN PYTHON

Python has a package called `re` [1] that allows you to search for regular expressions in strings.

---

[1] https://docs.python.org/3/library/re.html

The following snippet of code will allow you to find email addresses in a string:

```
import re
string1 = "example string with an email address: abc@xyz.edu"
match1 = re.search(r"([a-z]+)@([a-z]+).([a-z]+)", string1)
print(match1.group(0))  # prints abc@xyz.edu
```

The expression [a-z] represents all the letters a through z and the + indicates that there should be one or more of these characters.

The following table contains some of the special syntax for regular expressions in Python.

syntax	meaning	example	in words
[]	a set of characters	[a-zA-Z]	the letters a to z and A to Z
*	zero or more instances	[a-z]*	zero or more letters from a to z
+	one or more instances	[a-z]+	one or more letters from a to z
{}	exact number of instances	[a-z]{3}	exactly three letters from a to z
\d	a digit (0-9)	\d{3}	a number with exactly 3 digits
\w	the characters a-z, A-Z, 0-9, _	\w+	one or more characters from a-z, A-Z, 0-9, _

We can now use this to see how to extract the information we are looking for from the HTML example from earlier in this chapter.

The following snippet of code will allow us to search for Turing award winners:

```
import re
html ='<span>1966</span><a href="/perlis">Perlis, Alan J</a>'
pattern = r'<span>(\d{4})</span><a href="/([a-z]+)">([\w, ]+)</a>'
match2 = re.search(pattern, html)
print(match2.group(0)) # prints the whole string html
print(match2.group(1)) # prints 1966
print(match2.group(2)) # prints perlis
print(match2.group(3)) # prints Perlis, Alan J
```

Notice that the first expression (\d{4}) finds the year with exactly 4 digits, the second ([a-z]+) identifies the URL, and the third ([\w,]+) matches the letters, comma, and space in the name.

5.7 CHAPTER SUMMARY AND KEY CONCEPTS

- Regular expressions have three operations: **union** (+), **concatenation** (·), and **Kleene star** (*).

- Regular expressions are formally defined by these rules:

  - The empty set $\emptyset$ is a regular expression.
  - The empty string λ is a regular expression.
  - For any $s \in \Sigma$, s is a regular expression.
  - For any regular expressions r_1 and r_2, $r_1 + r_2$ is a regular expression.
  - For any regular expressions r_1 and r_2, $r_1 \cdot r_2$ is a regular expression.
  - For any regular expression r_1, $(r_1)^*$ is a regular expression.

- Regular expressions are used in some **text editors** and on the **Linux command line**.

- Regular expressions can be computed in programming languages such as **Python**.

EXERCISES

List the strings of length at most 4 for each of the following languages in shortlex order (see Section 1.10):

5.1 $00(0 + 1)^*$

5.2 $0(0 + 1)^*0$

5.3 $(0 + 1)^*0(0 + 1)^*1(0 + 1)^*$

5.4 $(0 + 1)^2 1(0 + 1)$

5.5 $(1 + 00)^*$

Give regular expressions for the following languages:

5.6 Every string over $\{0, 1\}$

$$\Sigma = \{a, b\}$$

5.7 Starts with b

5.8 Ends with ab

5.9 Contains at least one a

5.10 Contains the substring aab

5.11 Contains the characters a, a, b in that order but not necessarily next to one another

5.12 The 4th symbol is an a

5.13 The 4th last symbol is a b

5.14 Strings of length divisible by 3

5.15 Contains exactly one a

5.16 Contains exactly two bs

5.17 Contains at most two bs

5.18 Contains at least one a and one b (in any order)

5.19 Contains an even number of as and any number of b's

5.20 Does not contain the substring aa

5.21 Does not end ab

$$\Sigma = \{a - z, 0 - 9, \_\}$$

5.22 Valid variable name (cannot start with a digit, ignore reserved words)

5.23 Valid password (at least one letter and one digit)

$$\Sigma = \{0 - 9, \text{-}\}$$

5.24 Valid integer (no leading 0s, but could be negative)

5.25 Phone numbers of the form 555-555-5555

Write Python code to search strings for the following patterns:

5.26 Phone numbers of the form (555)-555-5555

5.27 Dates of the form MM-DD-YYYY (you don't have to check that the days or months are valid)

5.28 Email addresses of the form username@domain.com that could contain the symbols a-z, A-Z, 0-9, and at most one period.

5.29 Email addresses that may have multiple periods (e.g., username@domain.co.uk)

CHAPTER 6

Equivalence of Regular Languages and Regular Expressions

6.1   WHY YOU SHOULD CARE

In this section we will see that the languages accepted by DFAs and NFAs are exactly the same as the languages of regular expressions. This means that we can use these two ideas interchangeably. Moreover, we will prove the equivalence constructively, which means that we will have a way to design a regular expression from any DFA and vice versa.

6.2   CONVERTING A REGULAR EXPRESSION TO A λ-NFA

We will first see a procedure for converting any regular expression to a λ-NFA. This λ-NFA can then be converted to a DFA using the subset construction technique from Chapter 4. Recall from Section 5.4 that regular expressions are defined recursively using the following rules:

1. The empty set has the regular expression $\emptyset$.

2. The set $\{\lambda\}$ has the regular expression λ.

3. For each symbol $s \in \Sigma$, the set $\{s\}$ has the regular expression s.

4. For any regular expressions r_1 and r_2, the set $L(r_1) \cup L(r_2)$ has the regular expression $r_1 + r_2$.

5. For any regular expressions r_1 and r_2, the set $\{s_1 \cdot s_2 : s_1 \in L(r_1) \wedge s_2 \in L(r_2)\}$ has the regular expression $r_1 \cdot r_2$.

6. For any regular expression r_1, the set $\{s_1 \ldots s_n : n \geq 0 \wedge s_1, \ldots, s_n \in L(r_1)\}$ has the regular expression $(r_1)^*$.

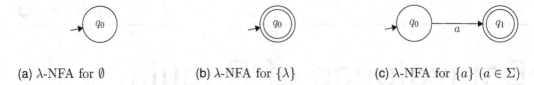

(a) λ-NFA for $\emptyset$ (b) λ-NFA for $\{\lambda\}$ (c) λ-NFA for $\{a\}$ $(a \in \Sigma)$

Figure 6.1 λ-NFAs for simple regular expressions

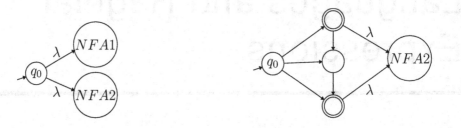

Figure 6.2 A λ-NFA for the union of the languages of two λ-NFAs

Figure 6.3 A λ-NFA for the concatenation of the languages of two λ-NFAs

The λ-NFAs for the languages $\emptyset$, $\{\lambda\}$, and $\{a\}$ for any $a \in \Sigma$ are given in Figures 6.1a, 6.1b, and 6.1c. Using these as our starting point, we see how to construct expressions that involve unions, concatenation, and Kleene *.

We can construct a λ-NFA for the union of two languages for which we have λ-NFAs by using λ-transitions from a new start state to the start states of the two λ-NFAs. Figure 6.2 shows a scheme for how you can do this for any pair of λ-NFAs.

We can concatenate the language of two λ-NFAs by adding λ-transitions from each of the accepting states of the first λ-NFA (and making them non-accepting) to the start state of the second λ-NFA. This is illustrated in the scheme shown in Figure 6.3.

Lastly, we can take the Kleene * of the language of any λ-NFA by using the scheme in Figure 6.4. We create a new start state (labeled q_0' in the figure), make it accepting, and then add a λ-transition to the start state of the original λ-NFA. We then add λ-transitions from each accepting state back to the start state q_0'.

We can create an NFA for any regular expression by repeating these three procedures as needed.

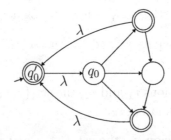

Figure 6.4 A λ-NFA for the Kleene * of the language of a λ-NFA

Example 6.1

We see below how a λ-NFA for the regular expression $ab(a + b)^*$ would be constructed. We start with the base λ-NFA for a (q_0 and q_1). To concatenate it with the base λ-NFA for b (q_2 and q_3), we add a transition from q_1 (the accepting state for the λ-NFA for a) to q_2 (the start state for the λ-NFA for b) and make q_1 a non-accepting state.

To get the expression $(a + b)$, we start with base λ-NFAs for a (q_6 and q_7) and b (q_8 and q_9) and create a new state (q_5) that goes to these λ-NFAs on a λ.

To get the expression $(a+b)^*$, we create a new state (q_4) and add a λ-transition from it to the start state of the λ-NFA for $(a+b)$ (q_5). We also add λ-transitions from the accepting states of the λ-NFA for $(a + b)$ (q_7 and q_9) to q_4.

Finally, to concatenate the λ-NFAs for ab and $(a + b)^*$, we add a λ-transition from the accepting state of the λ-NFA for ab (q_3) to the start state of the λ-NFA for $(a + b)^*$ (q_4) and make q_3 non-accepting.

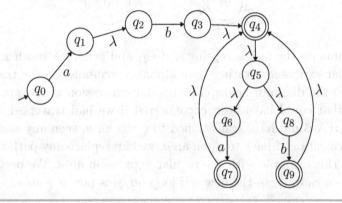

Example 6.2

A λ-NFA for the expression $(aa)^*(bc) + ab$ is given below:

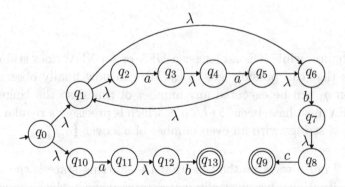

6.3 CONVERTING A DFA TO A REGULAR EXPRESSION

We can convert DFAs to regular expressions using a state elimination algorithm. This is useful in and of itself because there are some regular languages for which it is easy to define a DFA, but hard to come up with a regular expression. This algorithm then can be used to extract a regular expression from a DFA. Before formally defining the algorithm, let us see a few simple examples.

Consider the DFA below for accepting all the strings over $\{a, b\}$ that have an even number of a's.

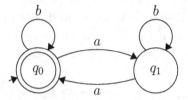

We can eliminate the non-accepting state q_1 and generate another figure that has a regular expression, rather than alphabet symbols, on the transition arc as shown in the diagram below. The regular expression must capture the set of strings that could have been encountered if we had traversed through q_1. Since the DFA could have transitioned to q_1 on an a, seen any number of b's, and then transitioned back to q_0 on an a, we can replace any path through the state q_1 in this example with the regular expression ab^*a. We previously had a self-loop on q_0 on b, so the new self-loop on q_0 is now $b + ab^*a$.

$$b + ab^*a$$

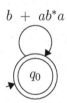

This new figure, while it is no longer a DFA or an NFA, does still capture all the strings that can possibly get to q_0. We can now finally observe that the self-loop on q_0 can be executed any number of times, so the language of the original DFA must have been $(b+ab^*a)^*$, which is precisely a regular expression for the set of strings with an even number of a's over $\{a, b\}$.

Now, it will not always be the case that the starting state is the only accepting one. In these situations, we must eliminate states down to just the start state and each accepting state. We see an example of this next.

Example 6.4

Consider the DFA below that accepts strings over $\{0,1\}$ that have an odd number of 0's and end in 1.

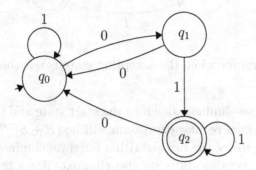

We eliminate the state q_1 since it is neither the start state nor an accepting state to get to the diagram below. To get this diagram, we have to consider all possible pairs of uneliminated states and replace the transitions from the first to the second that went directly through the eliminated state. In this example, since we could have gone from q_0 to q_1 to q_0 on 00, we add 00 to the self-loop on q_0. Similarly, we could have gone from q_0 to q_1 to q_2 on a 01, so we add a new transition from q_0 to q_2 on a 01. Note that there was no direct transition from q_2 to q_1, so we don't have to consider the possibilities of q_2 to q_0 through q_1 or q_2 to q_2 through q_1 here.

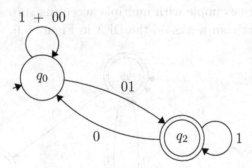

We can now generate a regular expression from this reduced figure as follows. Note that we can transition from q_0 to q_0 any number of times from a string in $1 + 00$ (because of the self-loop on q_0) or in 011^*0 (by transitioning to q_2 and then back to q_0). This can be represented by the regular expression $(1 + 00 + 011^*0)^*$. Any strings that are accepted would have to make one final transition to q_2 and potentially self-loop there, so the final regular expression in this instance would be $(1 + 00 + 011^*0)^*011^*$.

We can now generalize the above two examples for any DFA. If we have eliminated down to the start state and it is an accepting state with a self-loop with expression R, then the regular expression for the original DFA was $(R)^*$. If there is a non-starting

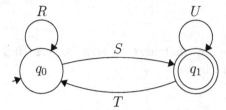

Figure 6.5 General form for when the accepting state is not the start state

accepting state, then we eliminate down to the start state and the accepting state as in Figure 6.5 and the final regular expression will be $(R + SU^*T)^*SU^*$.

We are now ready to see the full algorithm for state elimination (Algorithm 6.1). Note that if q_0 is an accepting state, we also eliminate down to just q_0 by itself.

Algorithm 6.1: StateElimination$(D = (Q, \Sigma, \delta, q_0, F))$

input : A DFA $D = (Q, \Sigma, \delta, q_0, F)$
output: A regular expression that has the same language as the DFA
1: **for all** $q_a \in F$ **do**
2:     Eliminate down to q_0 and q_a.
3:     Generate a regular expression for the strings that reach q_a.
4: Return the union of all the above regular expressions

Let us see one more example with multiple accepting states. We will see how the State Elimination algorithm works on the DFA in Figure 6.6.

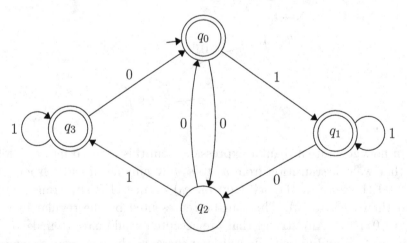

Figure 6.6 Example DFA for State Elimination algorithm

Example 6.5

This is the start of the State Elimination algorithm for the DFA in Figure 6.6. We need to eliminate down to q_0 and q_1, q_0 and q_3 and also q_0 by itself. In all cases, we need to eliminate q_2, so we do that first. We have to be careful to check for paths from each of the states q_0, q_1, and q_3 to each of the states q_0, q_1, and q_3 (including from a state to itself) to check for possible paths before eliminating q_2. In this case, the only possible paths were q_0 to q_0 (on 00), q_0 to q_3 (on 01), q_1 to q_0 (on 00), and q_1 to q_3 (on 01).

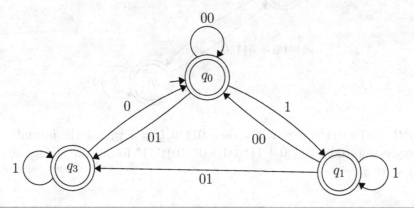

Example 6.6

Next, we eliminate state q_1 to get down to just states q_0 and q_3. Note that the only paths possible through q_1 were from q_0 to q_0 on 11*00 (the 1* coming from the self-loop on q_1) and q_0 to q_3 on 11*01.

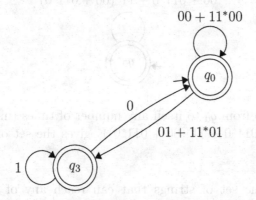

Using the formula from before with $R = 00 + 11^*00$, $S = 01 + 11^*01, T = 0, U = 1$, we get the expression $(00 + 11^*00 + (01 + 11^*01)1^*0)^*(01 + 11^*01)1^*$ for all strings that can reach the state q_3.

Example 6.7

Next, we revert to the diagram in Example 6.5 and eliminate down to q_0 and q_1 as seen below:

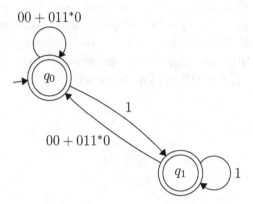

With $R = 00 + 011^*0, S = 1, T = 00 + 011^*0, U = 1$ we use the formula to get the expression $(00 + 011^*0 + 11^*(00 + 011^*0))^*11^*$ for all the strings that can reach state q_1.

Example 6.8

Lastly, we have to eliminate down to just the state q_0. We do so starting from Example 6.7. The only path we have to check is from q_0 to itself, which gives the expression $11^*(00 + 011^*0)$.

$$00 + 011^*0 + 11^*(00 + 011^*0)$$

Since we can loop from q_0 to itself any number of times (including zero), the expression $(00 + 011^*0 + 11^*(00 + 011^*0))^*$ gives the set of strings that can reach q_0.

Finally, to get the set of strings that can reach any of the accepting states $\{q_0, q_1, q_3\}$, we union the expressions from Examples 6.6, 6.7, and 6.8, to get the expression

$$(00 + 11^*00 + (01 + 11^*01)1^*0)^*(01 + 11^*01)1^* \quad +$$
$$(00 + 011^*0 + 11^*(00 + 011^*0))^*11^* \quad +$$
$$(00 + 011^*0 + 11^*(00 + 011^*0))^*.$$

While this formula is long and complicated, you can verify that it correctly captures the strings that get to each state. Here, q_0 is reachable by strings that have an even number of 0s and don't end in 1, q_1 is reachable by strings with an even number of 0s and do end in 1, and q_3 is reachable by strings that have an odd number of 0s and do end in 1.

6.4 ANOTHER DEFINITION FOR REGULAR LANGUAGES

Now that we have established that the language of every finite automaton has a regular expression and vice versa, we can expand our definition of regular languages even further.

> ### Definition 6.1
>
> A regular language is one that has a DFA, an NFA, or a regular expression.

6.5 CHAPTER SUMMARY AND KEY CONCEPTS

- There is a procedure for converting any regular expression into a λ-NFA.

- Any DFA can be converted to a regular expression using the **State Elimination** algorithm.

- A **regular language** is one that has a DFA, NFA, or regular expression.

EXERCISES

Give λ-NFAs *using the algorithm in this chapter* for the regular expressions below:

6.1 $ab(a + b)$

6.2 $(a + ba)^*$

6.3 $ab + (ab)^*$

6.4 $(a + b)^*(ab)(a + b)^*$

6.5 $(a + bb)^* + (b + aa)^*$

6.6 $(000)^*$

6.7 $00(0 + 1)^*00$

6.8 $0(0 + 1)^*0 + 1(0 + 1)^*1$

6.9 $(0 + 1)^*0(0 + 1)^2$

6.10 $001^*00 + 110^*11 + (01)^*(0 + 11)^*$

Construct DFAs for the following languages and *use the state elimination algorithm in this chapter* to create a regular expression for each. Show the intermediate steps as you eliminate each state along the way.

6.11 $\{w \in \{0,1\}^* : w \text{ is of even length}\}$

6.12 $\{w \in \{0,1\}^* : w \text{ has an odd number of 1s}\}$

6.13 $\{w \in \{0,1\}^* : w \text{ has at least two 1s}\}$

6.14 $\{w \in \{0,1\}^* : w \text{ has at most three 0s}\}$

6.15 $\{w \in \{0,1\}^* : \text{the number of 0s in } w \text{ is divisible by 3}\}$

6.16 $\{w \in \{0,1\}^* : w \text{ does not have 01 as a substring}\}$

6.17 $\{w \in \{0,1\}^* : w \text{ has an even number of 0s and an odd number of 1s}\}$

6.18 $\{w \in \{0,1\}^* : w \text{ has an even number of 0s and an even number of 1s}\}$

6.19 $\{w \in \{0,1\}^* : w \text{ has an odd number of 0s and an even number of 1s}\}$

6.20 $\{w \in \{0,1\}^* : w \text{ has an odd number of 0s and an odd number of 1s}\}$

6.21 $\{w \in \{0,1\}^* : \text{the number of 1s in } w \text{ is not divisible by 4}\}$

CHAPTER 7

Direct Proof and Closure Properties

7.1   WHY YOU SHOULD CARE

Proofs are the means by which we demonstrate that an algorithm does what we intend for it to do. It is usually impossible to test all possible inputs to an algorithm so we have to make a logical argument about its correctness. Making rigorous arguments then is a critical skill for a computer scientist.

7.2   TIPS FOR WRITING PROOFS

When writing proofs, here are a few rules of thumb to keep in mind:

1. Write in full, grammatical sentences. Don't forget punctuation, especially periods after mathematical expressions.

2. Write in paragraph form.

3. Show where the proof begins and ends very clearly.

4. Know your audience. This will tell you how formal you have to be.

5. Define variables before they are used.

6. Don't start a sentence with math.

7.3   THE IMPORTANCE OF DEFINITIONS

When making an argument, we have to start with some basic assumptions. This is particularly true in the case of **direct proofs**, where we prove statements such as "If P is true, then Q is true." For this kind of proof, we start by assuming that P is true and then make a sequence of logical arguments to show that Q then also must be true. To help us along the way, we need to make use of **definitions**. As an example, consider the following two definitions.

DOI: 10.1201/9781003383284-7

Definition 7.1

A string s has **even length** if there exists an integer $n \geq 0$ such that $|s| = 2n$.

Definition 7.2

A string s has **odd length** if there exists an integer $n \geq 0$ such that $|s| = 2n+1$.

We can now use these definitions to prove theorems such as the following.

Theorem 7.1

Concatenating an even length string and an odd length string gives an odd length string.

This theorem can be re-written as an if/then statement as follows.

Theorem 7.2

If x is an even length string and y is an odd length string, then xy is an odd length string.

To rigorously prove such a theorem, we need to use the above definitions to introduce notation for the lengths of the string to get us started. To show the result, we will need to use a definition again. The proof of the above theorem is given below:

Proof

Let x be an even length string and y be an odd length string.
By the definition of even length, we know that there exists some integer $n \geq 0$ such that $|x| = 2n$.
By the definition of odd length, we know that there exists some integer $m \geq 0$ such that $|y| = 2m + 1$. (Notice that we didn't use n again since the length of x and y may not be related.)
Then, the length of the string xy is just the sum of the lengths of x and y, or $2n + 2m + 1 = 2(n + m) + 1$.
Since we can write the length of xy as $2k + 1$ (where $k = n + m \geq 0$ is an integer), this means that xy is an odd length string.

Notice that we start by stating the assumption and defining our variables, even if they were defined in the theorem statement. This is always the first step in a direct proof. We proceed by unpacking the definition of our assumption, in this case introducing more notation to indicate that the strings are of even and odd lengths. At this point it is helpful to jump ahead to look at what we are trying to prove— here, that the concatenated string has odd length. We can work backwards from here by seeing that we will need to show using the definition of odd length that the concatenated string has odd length in the penultimate step. This gives us a direction for how to go about the proof: we must use the notation introduced by the definition of even and odd length strings to show that that concatenated string has an odd length. All that is left is to develop the reasoning to fill in the middle of this proof.

This proof follows the style conventions mentioned previously. We must clearly show where the proof begins by making the assumption and where it ends by stating the conclusion. Every step must be written out using complete sentences with punctuation, including periods ending sentences with math in them. All the variables should be defined before they are used and no sentence should start with a mathematical expression. Here is another similar example.

Theorem 7.3

If x, y, and z are all of odd length, then $xyyz$ has an even length.

Proof

Assume that x, y, and z are all strings of odd length.
By the definition of odd length, there exist integers $a, b, c \geq 0$ such that $|x| = 2a + 1$ and $|y| = 2b + 1$ and $|z| = 2c + 1$.
Then, the length of $xyyz$ is the sum of the lengths of its parts or $(2a + 1) + (2b + 1) + (2b + 1) + (2c + 1) = 2a + 4b + 2c + 4$.
Now, we can write this last expression as $d = 2(a + 2b + c + 2)$, where $(a + 2b + c + 2) \geq 0$ is an integer, so this means that $xyyz$ is of even length.

Sometimes a proof needs cases. Consider the following definition and theorem.

Definition 7.3

Two strings have the same parity if they are both even or both odd length.

Theorem 7.4

If two strings have the same parity, then their concatenation has even length.

Proof

Assume that two strings a and b have the same parity. There are two cases:
Case 1 (a and b have even lengths): We can write the length of a as $|a| = 2m$ and the length of b as $|b| = 2n$, for some integers $n, m \geq 0$. Then, $|ab| = 2m + 2n = 2(m + n)$, which means that ab is of even length since $(m + n) \geq 0$ is an integer.
Case 2 (a and b have odd lengths): We can write the length of a as $|a| = 2m+1$ and the length of b as $|b| = 2n + 1$, for some integers $n, m \geq 0$. Then, $|ab| = (2m + 1) + (2n + 1) = 2(m + n + 1)$, which means that ab is of even length since $(m + n + 1) \geq 0$ is an integer.
Since ab is of even length in all possible cases, the result follows.

For some direct proofs, it is easier to show the contrapositive. Recall from Chapter 3 that the contrapositive of the statement $A \Rightarrow B$ is the statement $\neg B \Rightarrow \neg A$, and they are logically equivalent. An example is given below:

Theorem 7.5

If xxx is an odd length string then x is an odd length string.

Proof

We can write out the statement "If xxx is an odd length string then x is an odd length string" as (xxx is odd length $\Rightarrow$ x is an odd length string). This means that the contrapositive of the statement is (x is NOT an odd length string $\Rightarrow$ xxx is NOT an odd length string), or "If x is not an odd length string, then xxx is not of odd length." Since another way of saying that a string is not of odd length is that it is of even length, we finally get the following equivalent statement: "If x is an even length string, then xxx is an even length string." This is easy to show using the direct proof technique we've seen so far.

Assume that x is an even length string. We can write the length of x as $|x| = 2n$ for some integer $n \geq 0$. Then, the length of xxx is $2n + 2n + 2n = 6n$, which means that xxx is an even length string since $6n$ can be written as $2(3n)$, where $3n \geq 0$ is an integer. Thus, we have shown the contrapositive of "If xxx is an odd length string then x is an odd length string" is true which means that the statement itself must be true.

We sometimes get an exact characterization of a property when a conditional statement and its converse is true. To show that $A \Leftrightarrow B$ we have to show both $A \Rightarrow B$ and $B \Rightarrow A$. This is called an "if and only if" (sometimes abbreviated iff) proof.

Theorem 7.6

The string xxx is an odd length string if and only if x is an odd length string.

Proof

To prove an if and only if statement we have to show both directions.

If x is an odd length string, then xxx is an odd length string: Assume that x is an odd length string. We can write the length of x as $|x| = 2n + 1$ for some integer $n \geq 0$. Then, the length of xxx is $3(2n + 1) = 6n + 3$, which means that xxx is an odd length string since $6n + 3$ can be written as $2(3n + 1) + 1$, and $(3n + 1) \geq 0$ is an integer.

If xxx is an odd length string, then x is an odd length string: See the proof of Theorem 7.5.

Since we have shown implications in both direction this means that the string xxx is an odd length string iff x is an odd length string.

7.4 NUMERICAL PROOFS

When comparing the running time of two algorithms we want to show that one is faster than the other. For example, we may have Algorithm A that on an input of size n needs $10n + 4$ operations, while Algorithm B needs n^2 operations. If we were to only look at small values of n, it would appear that Algorithm B is better. However, we may want to show that Algorithm A is preferable for larger values of n. We start by trying different values of n until we find one for which $10n + 4 < n^2$. After some trial and error we find that this inequality holds true when $n = 11$. But perhaps it is no longer true for some larger value (like $n = 100$)? We can show this is not the case using a direct proof.

Theorem 7.7

If $n \geq 11$, then $10n + 4 < n^2$.

Proof

Let us assume that $n \geq 11$. Then we can write

$$
\begin{aligned}
10n + 4 &< 10n + n \text{ (as } 4 < n) \\
&= 11n \\
&\leq n^2 \text{ (as } 11 \leq n),
\end{aligned}
$$

which shows that $10n + 4 < n^2$ because we have a string of (in)equalities and at least one of them was a strict inequality. Thus, $10n + 4$ is strictly smaller than n^2 whenever $n \geq 11$.

Here is another example.

Theorem 7.8

If $n \geq 5$, then $3n^2 + 10n \leq n^3$.

Proof

Let us assume that $n \geq 5$. Then we can write

$$
\begin{aligned}
3n^2 + 10n &\leq 3n^2 + 2n^2 \text{ (as } 10n \leq 2n^2 \text{ since } 5 \leq n) \\
&= 5n^2 \\
&\leq n^3 \text{ (as } 5 \leq n),
\end{aligned}
$$

which shows that $3n^2 + 10n \leq n^3$. This time the inequality is not strict because there were no strict inequalities. Thus, $3n^2 + 10n$ is at most n^3 whenever $n \geq 5$.

7.5   CLOSURE UNDER SET OPERATIONS

We can use direct proofs to show that performing a variety of operations on regular languages will result in languages that are also regular. This is useful as another way to show that a language is regular. For example, showing closure under complements easily lets us see that the set

$$\{s \in \{a, b\}^* : s \text{ doesn't start with an } a \text{ or } s \text{ doesn't end with a } b\}$$

is regular because the complement of this set is the language of the expression $a(a + b)^*b$, which we know is regular as it has a regular expression.

> **Theorem 7.9**
>
> If L_1 and L_2 are regular languages, then $L_1 \cup L_2$ is a regular language.

> **Proof**
>
> Let L_1 and L_2 be regular languages. Then there exist regular expressions r_1 and r_2 such that $L_1 = L(r_1)$ and $L_2 = L(r_2)$ by the equivalence of regular languages and regular expressions.
>
> Then, the regular expression $r_1 + r_2$ is a regular expression for the language $L_1 \cup L_2$. Since $L_1 \cup L_2$ has a regular expression, it must be regular.

A nearly identical proof shows closure under concatenation:

> **Theorem 7.10**
>
> If L_1 and L_2 are regular languages, then $\{xy : x \in L_1 \text{ and } y \in L_2\}$ is a regular language.

> **Proof**
>
> Let L_1 and L_2 be regular languages. Then there exists regular expressions r_1 and r_2 such that $L_1 = L(r_1)$ and $L_2 = L(r_2)$ by the equivalence of regular languages and regular expressions.
>
> Then, the regular expression $r_1 r_2$ is a regular expression for the language $\{xy : x \in L_1 \text{ and } y \in L_2\}$. Since $\{xy : x \in L_1 \text{ and } y \in L_2\}$ has a regular expression, it must be regular.

As practice, use the above technique to show that the following theorem is true:

> **Theorem 7.11**
>
> If L is a regular language, then
>
> $$\{x_1 x_2 x_3 \ldots x_n : n \geq 0 \text{ and } x_1, x_2, x_3, \ldots, x_n \in L\}$$
>
> is a regular language.

Sometimes a proof will require you to use a different version of a definition to show the result more easily. We know that there are three equivalent definitions for a regular language: the language must have a DFA, an NFA, or a regular expression. We take advantage of this in the following result:

Theorem 7.12

If L is a regular language, then $\overline{L}$ is a regular language.

Proof

Let L be any regular language. Then L must have a DFA, say $D = (Q, \Sigma, \delta, q_0, F)$. It is easy to see that we can create another DFA that accepts the strings rejected by this DFA and vice versa by switching accepting and rejecting states. In other words, the DFA $D' = (Q, \Sigma, \delta, q_0, Q - F)$ accepts $\overline{L}$. Therefore, $\overline{L}$ is regular as it has a DFA that accepts it.

Once we have proved some closure properties, we can even start to combine them together in our proofs. Consider the proof for closure under intersections below.

Theorem 7.13

If L_1 and L_2 are regular languages, then $L_1 \cap L_2$ is a regular language.

Proof

Let L_1 and L_2 be regular languages. To show that $L_1 \cap L_2$ is regular, we will use De Morgan's Law that says that $\overline{L_1 \cap L_2} = \overline{L_1} \cup \overline{L_2}$. Taking the complement of both sides, we get that $\overline{\overline{L_1 \cap L_2}} = \overline{\overline{L_1} \cup \overline{L_2}}$, which is the same as saying that $L_1 \cap L_2 = \overline{\overline{L_1} \cup \overline{L_2}}$ since the complement of the complement of a set is just the original set. We are now ready to prove the result.

Since L_1 and L_2 are regular, by the closure under complements property we know that $\overline{L_1}$ and $\overline{L_2}$ are regular. From the closure under union property, this tells us that $\overline{L_1} \cup \overline{L_2}$ is regular. Again using the closure under complements property, we get that $\overline{\overline{L_1} \cup \overline{L_2}}$ is regular. But this is the same as $L_1 \cap L_2$ as we showed above. Thus, $L_1 \cap L_2$ is regular.

Here is another example where we can combine several properties together to show closure under set difference.

Theorem 7.14

If L_1 and L_2 are regular languages, then $L_1 - L_2$ is a regular language.

Proof

Let L_1 and L_2 be regular languages. By closure under complements, we know that $\overline{L_2}$ must be regular. Furthermore, by closure under intersections we know that $L_1 \cap \overline{L_2}$ is regular. But $L_1 \cap \overline{L_2}$ is the same as $L_1 - L_2$ (for example, compare the Venn diagrams for both), so $L_1 - L_2$ is regular.

We can also show closure under the reverse operation. We define w^R to be the reverse of any string w.

Example 7.1

For example, $1100^R = 0011$, $010101^R = 101010$, $abbbab^R = babbba$, $a^R = a$, and $\lambda^R = \lambda$.

We define the reverse of a language to be the set of strings that are the reverse of the strings in the language.

Definition 7.4

The reverse of a language L is $L^R = \{s^R : s \in L\}$.

Theorem 7.15

If L is a regular language, then L^R is a regular language.

Rather than formally proving this claim, we will see a construction that demonstrates it for any regular language. We'll see a formal proof of the claim in Chapter 8.

Any regular language L must have a regular expression E. We will see how to take the regular expression and convert it into a regular expression E^R for the language L^R. Recall that regular expressions were defined recursively using six rules in Section 5.4. We show how to reverse the expression for each of these rules as follows:

1. We leave instances of $\emptyset$, λ, and single characters $s \in \Sigma$ unchanged.

2. If the expression is of the form $E_1 + E_2$, we replace it with the expression $E_1^R + E_2^R$ (recursively following this procedure for the expressions E_1 and E_2).

3. If the expression is of the form $E_1 \cdot E_2$, we replace it with the expression $E_2^R \cdot E_1^R$.

4. If the expression is of the form $(E_1)^*$, we replace it with the expression $(E_1^R)^*$.

Example 7.2

To get the reverse of the regular expression $01^* + 10^*$, we compute it as

$$
\begin{aligned}
(01^* + 10^*)^R &= (01^*)^R + (10^*)^R \text{ (using rule 2)} \\
&= (1^*)^R 0^R + (0^*)^R 1^R \text{ (using rule 3 twice)} \\
&= (1^*)^R 0 + (0^*)^R 1 \text{ (using rule 1 twice)} \\
&= (1^R)^* 0 + (0^R)^* 1 \text{ (using rule 4 twice)} \\
&= 1^* 0 + 0^* 1 \text{ (using rule 1 twice)}.
\end{aligned}
$$

Similarly, $(ab^*)^R = b^* a$, $(0(01)^*1)^R = 1(10)^*0$, $(a + ba^*)^R = a + a^* b$, $(ab(a + b)^* + (a + b)^* aa)^R = (a + b)^* ba + aa(a + b)^*$, and $(ab^* ab^* + (a + b)b^* aa)^R = b^* ab^* a + aab^*(a + b)$.

While we can verify that the reverse of expressions seems to work for all the examples (and perhaps several more that you have tried), this doesn't definitively prove that this procedure always works. We should be skeptical when something works for a few cases because it may stop working for a more complicated case. You will need to learn a proof technique called strong induction later in this book to show that this procedure does correctly reverse any expression to give the reverse of its language.

7.6 CHAPTER SUMMARY AND KEY CONCEPTS

- Proofs are how computer scientists demonstrate the **correctness and efficiency** of their algorithms.

- A proof is written for another person. Know who is your **audience** and make sure the proof is easy for them to read.

- Knowing how to use **definitions** is central to understanding how to start and how to end a proof.

- **Closure properties** give us an easy way to show that a language is regular.

- Regular languages are **closed** under union, concatenation, Kleene star, complement, intersection, difference, and reversal.

EXERCISES

7.1 Show that the concatenation of two even length strings gives an even length string.

7.2 Show that if x and y are odd length strings and z is an even length string that xyz is an even length string.

7.3 Show that if x and y are odd length strings and z is an even length string that $xyyz$ is an odd length string.

7.4 Show that if for strings x, y, z if xy and xz are both odd length, then yz must be even length.

7.5 Show that for all $n \geq 8$, $7n + 100 < 20n$.

7.6 Show that for all $n \geq 101$, $100n + 100 < n^2$.

7.7 Show that for all $n \geq 501$, $1000n + 1000 < 2n^2$.

7.8 Show that for all $n \geq 3$, $4n^2 + 6n \leq 2n^3$.

7.9 Show that if L_1, L_2, and L_3 are regular languages that $L_1 \cup L_2 \cup L_3$ is also a regular language.

7.10 Define the NOR operation as $NOR(L_1, L_2) = \{x : x \notin L_1 \wedge x \notin L_2\}$. Show that regular languages are closed under the NOR operator.

7.11 The symmetric difference of two sets is the set of elements that are in exactly one of the two sets. Show that the regular languages are closed under symmetric difference.

7.12 Show that if L_1, L_2, and L_3 are regular languages that $L_1 \cap L_2 \cap L_3$ is also a regular language.

7.13 Prove that any language with exactly two strings must be regular.

7.14 Prove that any finite language must be regular.

CHAPTER 8

Induction

8.1 WHY YOU SHOULD CARE

Proof by induction is a key proof technique that is used extensively in computer science. It is used for proving correctness of algorithms with loops and recursion. It is also used for showing bounds on the run-time of many recursive algorithms. We'll also see how some of the results for sets and regular languages from earlier chapters can be formally proved.

8.2 INDUCTION AND RECURSION

Consider the pseudo-code in Algorithm 8.1 for computing the factorial function.

Algorithm 8.1: factorial(n)

input : An integer $n \geq 1$
output: The factorial of n ($n!$), defined as $1 \times 2 \times 3 \times \ldots \times n$
  1: **if** $n == 1$ **then**
  2:     **return** 1
  3: **return** factorial($n - 1$) * n

Why does the above algorithm work? How do we go about proving that it always returns the correct answer? The intuitive reasoning goes something like this:

1. When $n = 1$ (the smallest possible input), the function returns the correct answer of 1 in the second line.

2. When $n = 2$, the third line of the function returns factorial(1) * 2, which is 1 * 2 or 2, again the correct answer.

3. When $n = 3$, the third line of the function returns factorial(2) * 3, which is 2 * 3 or 6, again the correct answer.

4. When $n = 4$, the third line of the function returns factorial(3) * 4, which is 6 * 4 or 24, again the correct answer.

5. This pattern can be continued indefinitely until we reach any desired positive integer.

The idea behind proof by induction is to formalize this pattern. We have to show that the function gives the correct answer for the smallest input $n = 1$ (called the base step). We then have to show that if the function works for the case of some $n = k$, then it also must work for the case when $n = k + 1$ (called the inductive step).

Let us use the above example to write a more formal proof. A proof by induction always has four parts and you should practice writing all four parts until you become a more experienced proof-writer. These parts are:

1. State the hypothesis

2. Base step

3. Inductive step

4. Conclusion

For ease of writing our proof, we denote the hypothesis by the function P to refer to it more succinctly later. The hypothesis is always a logical statement (i.e., taking on a value of true or false). A common error is to mistakenly think of $P(n)$ as a number, which it is not.

Example 8.1

Hypothesis: For all $n \geq 1$, $P(n)$: The factorial program on input n returns $1 \times 2 \times 3 \times \ldots \times n$.

The base step is usually the simplest part of a proof by induction. You should make sure to verify the hypothesis for the smallest possible value of n. This may not always be 1, depending on the result being proved.

Example 8.2

Base step: When $n = 1$, the first line of the algorithm returns the value 1. Since the factorial of 1 is also 1, the algorithm is correct in this case, and so $P(1)$ is true.

The inductive step of the proof is where all the "work" is done. It always takes the same form: fix k, assume the hypothesis for k, prove the hypothesis for $k + 1$ while making sure to use the assumption. More specifically, before writing the induction step, it helps to write down the inductive hypothesis ($P(k)$) and what you are trying to prove ($P(k + 1)$). Here, $P(k)$ is the statement "The factorial program on input k returns $1 \times 2 \times 3 \times \ldots \times k$" and $P(k + 1)$ is the statement "The factorial program on input $k + 1$ returns $1 \times 2 \times 3 \times \ldots \times (k + 1)$." We have to explain why it is that $P(k)$ implies $P(k + 1)$ for any $k \geq 1$.

Example 8.3

Induction step: Fix some $k \geq 1$. Assume that the hypothesis P holds true for $n = k$. We will show that it also holds true when $n = k + 1$.

Since the hypothesis holds true for $n = k$, we know that factorial(k) will return $1 \times 2 \times 3 \times \ldots \times k$. When the algorithm is run with an input of $k + 1$, we know that $k + 1 > 1$ (since $k \geq 1$) so it gets to the third line and returns factorial($n - 1$) * n which is factorial(k) * $(k + 1)$ as $n = k + 1$. From our inductive assumption above, this is the same as $1 \times 2 \times 3 \times \ldots \times k$ times $(k+1)$ or $1 \times 2 \times 3 \times \ldots \times (k+1)$, or $(k+1)!$, as desired. Hence, the hypothesis $P(k+1)$ is true as well.

The conclusion simply summarizes the proof by induction argument. It is somewhat verbose and is almost identical in all proofs, but I recommend that you write it out after each proof to remind yourself of why it is that induction works.

Example 8.4

Conclusion: We know that $P(1)$ is true from the base case. Since we know that $P(1)$ is true and we showed that $P(1)$ implies $P(2)$ in the inductive step (when $k = 1$), this means that $P(2)$ is true. Since we know that $P(2)$ is true and we showed that $P(2)$ implies $P(3)$ in the inductive step (when $k = 2$), this means that $P(3)$ is true. Continuing in this manner, we can show that $P(n)$ is true for all $n \geq 1$.

A proof doesn't just convince us that an algorithm is correct, it gives us a deeper understanding of *why* the algorithm is correct. It takes away the mystery of how the algorithm works and gives us insight into how to adapt it to solve new problems. Hopefully, the above proof gave you a deeper understanding and appreciation for the recursive factorial function.

8.3   AN ANALOGY FOR UNDERSTANDING INDUCTION

An analogy for induction is building a multi-floor building or tower. When constructing the building, it is important to make sure that the bottom-most level has a firm foundation, which is analogous to the base case of induction. We can only build the second level of the building after the first (the base) has been firmly established. Similarly, we need the second level before we can build the third level, and so on.

In an induction proof, we have to show the following two facts:

1. The bottom level or base case is firmly established. [Show $P(1)$ is true.]

2. If any level k is firmly established, then the level $k + 1$ will also be firmly established. [For all $k \geq 1$, show that if $P(k)$ is true then $P(k + 1)$ is true.]

Using these two facts, it is clear to see that we can build a tower of any height. The base level is firmly established by the first fact. The second level is firmly established

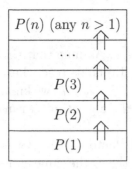

Figure 8.1 The tower analogy for induction

by applying the second fact with $k = 1$ to the base case. The third level is firmly established by applying the second fact with $k = 2$ to the previous result about the second level. Continuing in this way, we can show that any level $n \geq 1$ will be firmly established. This is illustrated in Figure 8.1.

8.4 INDUCTION FOR ANALYZING SORTING RUN-TIME

When analyzing the running time of sorting algorithms such as selection sort and insertion sort, it can be shown that to sort a list of n items takes time proportional to $1 + 2 + 3 + \ldots + (n - 1)$ time steps. To demonstrate that these sorting algorithms take quadratic time, it helps to show that $1 + 2 + 3 + \ldots + (n - 1) = n(n - 1)/2$ for all $n \geq 1$. Before proving this result, we will take a short detour to introduce some notation that you will encounter frequently in this book and in later courses.

Rather than write out long series such as $1 + 2 + 3 + \ldots + (n - 1)$, we can use summation or sigma notation (for the Greek letter used to represent it), as follows:

$$\sum_{i=1}^{n-1} i = 1 + 2 + 3 + \ldots + (n - 1).$$

In this notation, the Σ indicates summation (not to be confused with a string alphabet). The subscript (here, $i = 1$) indicates what value we start i at in the summation, the superscript (here, $n - 1$) indicates the (inclusive) stopping point, and the expression right after is the function that is applied to each of the values of i that are being summed. In this example, we are just adding up all these values of i.

Definition 8.1

For any integers $a, b \in \mathbb{Z}, a \leq b$ and function on integers f, the notation

$$\sum_{i=a}^{b} f(i)$$

is a shortened way of expressing the sum $f(a) + f(a + 1) + f(a + 2) + \ldots + f(b)$.

Example 8.5

Here are some more examples of this notation:

$$\sum_{i=1}^{4} i^2 = 1^2 + 2^2 + 3^2 + 4^2,$$

$$\sum_{i=0}^{n} 2^i = 2^0 + 2^1 + 2^2 + \ldots + 2^n,$$

$$\sum_{i=1}^{n} i(i+1) = 1 \cdot 2 + 2 \cdot 3 + 3 \cdot 4 + \ldots + n \cdot (n+1).$$

We can write infinite sums in this way:

$$\sum_{i=1}^{\infty} 2^{-i} = \frac{1}{2} + \frac{1}{4} + \frac{1}{8} + \ldots.$$

We can write nested summations (summations inside summations) in the same way loops can be nested:

$$\sum_{i=1}^{2} \sum_{j=2}^{3} ij^2 = 1 \cdot 2^2 + 1 \cdot 3^2 + 2 \cdot 2^2 + 2 \cdot 3^2.$$

Summations can also be computed over the elements of a set:

$$\sum_{x \in \{2,3,5\}} x^3 = 2^3 + 3^3 + 5^3.$$

We will now demonstrate $\sum_{i=1}^{n-1} i = n(n-1)/2$ using induction.

Example 8.6

Hypothesis: For all $n \geq 1$, $P(n)$: The sum $\sum_{i=1}^{n-1} i$ is equal to $n(n-1)/2$.

Once again, note that $P(n)$ is a logical statement (taking values true or false) and not the sum of the series. Specifically, it denotes whether or not the sum of the series is equal to the formula $n(n-1)/2$.

Example 8.7

Base step: When $n = 1$, the sum has zero terms and hence adds up to zero. The formula $n(n-1)/2 = 1(0)/2$ is also equal to zero. Hence the base case $P(1)$ holds.

Example 8.8

Induction step: Fix some $k \geq 1$. Assume that the hypothesis P holds true for $n = k$. We will show that it also holds true when $n = k + 1$.

Taking the left-hand side of the proposition for $n = k + 1$, we get

$$
\begin{aligned}
\sum_{i=1}^{n-1} i &= \sum_{i=1}^{(k+1)-1} i \\
&= \sum_{i=1}^{k} i \\
&= \left(\sum_{i=1}^{k-1} i \right) + k \\
&= k(k-1)/2 + k \qquad \text{(by the induction hypothesis)} \\
&= (k^2 - k)/2 + k \\
&= (k^2 - k + 2k)/2 \\
&= (k^2 + k)/2 \\
&= k(k+1)/2,
\end{aligned}
$$

which is equal to the formula for the right-hand side of the proposition for $n = k + 1$. Hence, the proposition holds true for $n = k + 1$ as well.

Let us understand how the derivation above was developed. First, start with what you are assuming $(P(k))$, which is $\sum_{i=1}^{k-1} i = k(k-1)/2$ and what you are trying to prove $(P(k+1))$, which is $\sum_{i=1}^{(k+1)-1} i = (k+1)((k+1)-1)/2$ or $\sum_{i=1}^{k} i = (k+1)k/2$. When proving equality, we start with one side of what we are trying to prove (here, $\sum_{i=1}^{k} i$) and we want to show that it is equal to the right side $((k+1)k/2)$. To do this, we have to make a connection to the induction hypothesis. For proving the sum of series, this will almost always look the same: observe that the sum of the larger series is the same as the smaller one with an additional term. This is why, for this example, we break down the summation $\sum_{i=1}^{k} i$ into $\sum_{i=1}^{k-1} i$ and k on the third line of the derivation. We can now apply the induction hypothesis to replace $\sum_{i=1}^{k-1} i$ with $k(k-1)/2$ on the fourth line. Finally, we use algebra to show that we get the expression on the right-hand side of the equality we are trying to prove $((k+1)k/2)$.

The conclusion step is identical to the one in the first example.

Example 8.9

Conclusion: We know that $P(1)$ is true from the base case. Since $P(1)$ is true and we showed that $P(1)$ implies $P(2)$ in the inductive step (when $k = 1$), this means that $P(2)$ is true. Since $P(2)$ is true and we showed that $P(2)$ implies $P(3)$ in the inductive step (when $k = 2$), this means that $P(3)$ is true. Continuing in this manner, we can show that $P(n)$ is true for all $n \geq 1$.

8.5 HOW MANY BIT STRINGS ARE THERE OF LENGTH (AT MOST) N?

We'll next use induction to prove a result about the number of distinct bit strings, that is, strings of length n over the alphabet $\Sigma = \{0, 1\}$. There is exactly one string of length 0, namely λ. For length one, we have two: $\{0, 1\}$. For length two we have 4: $\{00, 01, 10, 11\}$, and for length three we have eight: $\{000, 001, 010, 011, 100, 101, 110, 111\}$. Before reading ahead, what is the pattern that you notice?

You may have noticed that the number of strings of each successive length is twice that of the length one smaller. We might then use this to make the hypothesis $P(n)$: the number of strings of length n is equal to 2^n. Let us see how to prove this.

Theorem 8.1

For each $n \geq 0$, there are exactly 2^n strings of length n over $\{0, 1\}$.

Example 8.10

Hypothesis: For all $n \geq 0$, $P(n)$: There are exactly 2^n strings of length n over $\{0, 1\}$.

Base step: When $n = 0$, there is only one string, namely λ, and $2^0 = 1$, so $P(0)$ is true.

Inductive step: Fix some $k \geq 0$. Assume that the hypothesis P holds true for $n = k$. That is, there are exactly 2^k strings of length k over $\{0, 1\}$. Each string of length $k + 1$ must start with a 0 or a 1. The number of strings of length $k + 1$ that start with a 0 is equal to the number of strings of length k, since each length k string is possible as a suffix, for a total of 2^k such strings by the induction hypothesis. Similarly, the number of strings of length $k + 1$ that start with a 1 is equal to 2^k by the induction hypothesis. Thus, in total the number of strings of length $k + 1$ is equal to the number of such strings that start with 0 plus the number of such strings that start with 1, which is exactly $2^k + 2^k = 2 \cdot 2^k = 2^{k+1}$. Hence, $P(k + 1)$ is true as well.

Conclusion: We know that $P(0)$ is true from the base case. Since we know that $P(0)$ is true and we showed that $P(0)$ implies $P(1)$ in the inductive step (when $k = 0$), this means that $P(1)$ is true. Since we know that $P(1)$ is true and we showed that $P(1)$ implies $P(2)$ in the inductive step (when $k = 1$), this means that $P(2)$ is true. Continuing in this manner, we can show that $P(n)$ is true for all $n \geq 0$.

Notice that this case was slightly different than the previous examples as we started from $n = 0$. The induction step also needed a slightly more sophisticated argument showing that $P(k)$ implies $P(k + 1)$. The approach we used earlier still works. Start by writing out the statements for $P(k)$ and $P(k+1)$ fully for the problem. Then articulate why $P(k + 1)$ is true given the fact that $P(k)$ is true.

We can take this analysis one step further to answer the question of how many strings there are of length *at most* n. We know from the previous theorem that the number of strings of length n is exactly 2^n, so the number of strings of length at most n must be the number of strings of lengths $0, 1, 2, \ldots, n$, or $\sum_{i=0}^{n} 2^i$. When $n = 0$, this is just 2^0 or 1. When $n = 1$, we get $2^0 + 2^1 = 3$. When $n = 2$, we have $2^0 + 2^1 + 2^2 = 7$. When $n = 3$, it is $2^0 + 2^1 + 2^2 + 2^3 = 15$. Do you see a pattern emerging? Try to guess the pattern before reading on.

You may have noticed that the nth number consistently turns out to be $2^{n+1} - 1$. (This is not the only way of expressing the pattern; you might have also observed that each number is one more than twice the previous, which is also correct.) Let us state this formally as a hypothesis and see how we can prove it.

Theorem 8.2

For all $n \geq 0$, there are exactly $2^{n+1} - 1$ strings of length at most n over $\{0, 1\}$.

Example 8.11

By Theorem 8.1, there are exactly 2^n strings of length n over $\{0, 1\}$. Therefore the number of strings of length at most n over $\{0, 1\}$ will be equal to $\sum_{i=0}^{n} 2^i$. We will show that $\sum_{i=0}^{n} 2^i$ is equal to $2^{n+1} - 1$ via induction on n.

Hypothesis: For all $n \geq 0$, $P(n)$: $\sum_{i=0}^{n} 2^i = 2^{n+1} - 1$.

Base step: When $n = 0$, there is only one string of length at most 0, namely λ, and $2^{0+1} - 1 = 2 - 1 = 1$, so $P(0)$ is true.

Inductive step: Fix some $k \geq 0$. Assume that the hypothesis P holds true for $n = k$. That is, $\sum_{i=0}^{k} 2^k = 2^{k+1} - 1$. Taking the left-hand side of the proposition for $n = k + 1$, we get

$$
\begin{aligned}
\sum_{i=0}^{k+1} 2^i &= \left(\sum_{i=0}^{k} 2^i \right) + 2^{k+1} \\
&= (2^{k+1} - 1) + 2^{k+1} \quad \text{(by the induction hypothesis)} \\
&= 2^{k+2} - 1,
\end{aligned}
$$

which is equal to the formula for the right-hand side of the proposition for $n = k + 1$. Hence, the proposition holds true for $n = k + 1$ as well.

Conclusion: We know that $P(0)$ is true from the base case. Since $P(0)$ is true and we showed that $P(0)$ implies $P(1)$ in the inductive step (when $k = 0$), this means that $P(1)$ is true. Since $P(1)$ is true and we showed that $P(1)$ implies $P(2)$ in the inductive step (when $k = 1$), this means that $P(2)$ is true. Continuing in this manner, we can show that $P(n)$ is true for all $n \geq 0$. Since we showed earlier that the number of strings of length n over $\{0, 1\}$ is $\sum_{i=0}^{n} 2^i$, the above induction result shows that this is equal to $2^{n+1} - 1$.

Notice that this time we used the proof by induction as part of a larger proof. It is important to tell your reader that you are doing this, both in the setup to say how you will use the inductive hypothesis and in the conclusion explaining how you used the induction result for the larger proof.

TABLE 8.1 Table for 2^n and $n!$ for small values of n

n	1	2	3	4	5	6	7
2^n	2	4	8	16	32	64	128
$n!$	1	2	6	24	120	720	5040

8.6 COMPARING GROWTH OF FUNCTIONS

When you later learn about running times of algorithms, you will have to compare the running time of various functions to see which ones grow quicker. We will show that the factorial function ($n!$, seen earlier in this chapter) grows faster than an exponential function (2^n). Before trying to prove this result, we should check it for small values of n to make sure that it checks out. From Table 8.1 we can see that for small values of n ($n \leq 3$), the factorial function has a smaller value than this exponential one but for larger values ($n \geq 4$) the factorial function seems to get substantially larger with larger values of n.

The table is insufficient to tell us that the inequality holds true for all bigger values, so we will prove it by induction. Note that the base and conclusion steps will be different from the earlier examples as the hypothesis will hold only for $n \geq 4$.

Theorem 8.3

For all $n \geq 4$, $2^n < n!$.

Proof

Hypothesis: Let $P(n)$ be the logical statement $2^n < n!$ for any $n \in \mathbb{N}$.

Base step: When $n = 4$, we have that $2^n = 16$ and $n! = 24$, so $2^n < n!$ is true and the base case $P(4)$ holds.

Inductive step: Fix some $k \geq 4$. Assume that the hypothesis holds for $n = k$. This means that $2^k < k!$. To prove the hypothesis for $n = k + 1$, we show that the left-hand side of $P(k+1)$

$$
\begin{aligned}
2^{k+1} &= 2 \times 2^k \\
&< 2 \times k! \quad \text{(by the induction hypothesis)} \\
&< (k+1) \times k! \quad \text{(since } 2 < (k+1) \text{ for } k \geq 4) \\
&= (k+1)!.
\end{aligned}
$$

Since the string of (in)equalities includes one that is a strict inequality ($<$), this means that $2^{k+1} < (k+1)!$. Hence, the hypothesis is true for $n = k + 1$.

Conclusion: We know that $P(4)$ is true from the base case. Since $P(4)$ is true and we showed that $P(4)$ implies $P(5)$ in the inductive step (when $k = 4$), this means that $P(5)$ is true. Since $P(5)$ is true and we showed that $P(5)$ implies $P(6)$ in the inductive step (when $k = 5$), this means that $P(6)$ is true. Continuing in this manner, we can show that $P(n)$ is true for all $n \geq 4$.

While the above induction step is correct, it doesn't give a lot of insight of how it was obtained, so we'll break it down next. When proving the induction step, it is helpful to write down your assumption $P(k)$ (here, that $2^k < k!$) and what you are trying to prove (here, $2^{k+1} < (k+1)!$) first. This will help you to figure out the connection between them. In this case, we start with one side of the inequality we are trying to prove (here, 2^{k+1}) and figure out how it relates to our induction hypothesis ($2^k < k!$). We can make the observation that 2^{k+1} is just 2 times 2^k to make that connection, as we do in the second line of the derivation. We then use the induction hypothesis to replace 2×2^k with $2 \times k!$ in the third line, stating that we used the induction hypothesis. You should always make sure that you used the induction hypothesis somewhere in the inductive step. Next, we have to make sure that the value $2 \times k!$ is at most the value $(k+1)!$ from what we are trying to prove. To make this connection, we observe that $(k+1)!$ is just $(k+1) \times k!$ and so we really need to just show that $2 \times k!$ is at most $(k+1) \times k!$, which it must be as 2 is less than $(k+1)$ as we assumed that $k \geq 4$. This gives us the third line of the derivation above.

Notice how this proof is different from the previous ones in that our base case must be when $n = 4$ since $P(1), P(2), P(3)$ are all false. This is a good reminder that $P(n)$ may not always be true for all values of n. It is also a good cautionary example for why we must pay attention to the base case. The above induction step seems to work whenever $(k+1) > 2$ or $k > 1$, but the hypotheses $P(2)$ and $P(3)$ are false, so we really needed to check to make sure that the base case was true.

8.7 COMMON ERRORS WHEN USING INDUCTION

1. Not clearly defining the induction hypothesis: Any induction proof should start with a clear statement of the induction hypothesis $P(n)$ and for which values it is being proved to be true.

2. Treating $P(n)$ as a number: A common error, particularly in proofs involving summations, is to treat $P(n)$ as a number. Remember that $P(n)$ is a logical statement and can only take on the values true and false.

3. Omitting the base step: If we don't check the base case, we can run into serious issues. For example, consider the following untrue claim:

$$P(n) : 3^n \text{ is even, for all } n \geq 1$$

Proof (inductive step): Fix some $k \geq 1$. We assume the hypothesis is true for $n = k$, that is 3^k is even. This means that we can write $3^k = 2f$ for some $f \geq 1$. Then, 3^{k+1} can be written as $3^{k+1} = 3^k \cdot 3 = 2f \cdot 3$, which means that 3^{k+1} is even as it can be written as $2(3f)$.

This proof must clearly be erroneous since we know that no natural number power of 3 is even. This proof breaks down because the base step does not hold—when $n = 1$, $3^n = 3$ is not an even number.

4. Making an error in the inductive step: When proving the inductive step, you have to make sure that the proof works for all stated values. The following is an example of an incorrect proof that all horses are the same color!

$P(n)$: every horse in a set of n horses has the same color, for all $n \geq 1$

Base step: When $n = 1$, there is only one horse in the set and thus every horse in that set has the same color.

Inductive step: Fix some $k \geq 1$. We assume the hypothesis is true for $n = k$. When $n = k + 1$ we exclude the first horse and get a set of size k for which we know that all the horses are the same color (by the induction hypothesis). Again, doing the same but this time excluding the last horse, we get that all the horses in this set are the same color. Now, since both sets have horses of the same color, it follows that all the $k + 1$ horses are the same color.

The error here stems from the fact that the inductive step does not work when showing that $P(2)$ is true. The last statement in the proof is false.

8.8 STRONG INDUCTION

Sometimes, regular induction is hard to make work because we cannot build the argument based on one smaller value of n. In such cases, we need a stronger hypothesis—that the induction hypothesis is true for all smaller values of n. In such situations, we use a technique known as strong induction. The proof of the recursive Fibonacci program in Algorithm 8.2 below is a great example of this.

Algorithm 8.2: Fibonacci(n)

input : An integer $n \geq 1$
output: The nth Fibonacci number
1: **if** $n == 1$ or $n == 2$ **then**
2:    return 1
3: return Fibonacci($n - 1$) + Fibonacci($n - 2$)

Recall that the Fibonacci numbers start $1, 1, 2, 3, 5, \ldots$ and each subsequent one is the sum of the previous two. It is not obvious how to write a proof of correctness for the above algorithm by regular induction because computing the $(n + 1)$th Fibonacci number correctly relies on having correctly computed the nth and $(n-1)$th Fibonacci numbers.

We'll next see an example of strong induction that gets around this. Specifically, in strong induction we assume the hypothesis is true for all values up to n and then prove it for the case of $(n + 1)$. The other changes we make include showing multiple base cases and modifying the concluding step.

Theorem 8.4

The Fibonacci program on input n returns the nth Fibonacci number for all $n \geq 1$.

Proof

Hypothesis: $P(n)$: The Fibonacci program on input n returns the nth Fibonacci number.

Base step: When $n = 1$ and $n = 2$, the algorithm returns 1 and the first two Fibonacci numbers are both 1. So, $P(1)$ and $P(2)$ hold true.

Induction step: Fix some $k \geq 2$. Assume that the hypothesis holds for all $n \leq k$. In particular, this means that $P(k)$ and $P(k-1)$ must be true, or that the algorithm correctly computes the kth and $(k-1)$th Fibonacci numbers. Since the $(k+1)$th Fibonacci number is simply defined as the sum of the previous two, and this is what is returned on Line 3 as the function works correctly when $n = k$ and $n = k - 1$, we know that the algorithm returns the correct value when $n = k + 1$. Thus, the hypothesis is true for $n = k + 1$.

Conclusion: We know that $P(1)$ and $P(2)$ are true from the base case. Since we know that $P(1)$ and $P(2)$ are true, by the induction step this means that $P(3)$ is true. Since we know that $P(2), P(3)$ are true, this means that $P(4)$ is true. Continuing in this manner, we can show that $P(n)$ is true for all $n \geq 1$.

A question you might have about strong induction is how many base cases you need to show. This depends on the largest value of n that cannot be proved using the inductive argument. In this example, we cannot argue that the two smallest Fibonacci numbers were computed from any smaller numbers, so we had to include base cases for both. More generally, if your inductive step relies on the problem being a certain size, then you need to show the hypothesis for all smaller sizes in the base step.

One might infer from the name that strong induction is stronger than regular induction—you can prove more with it. This turns out not to be the case as they are equivalent in proving power. Any strong induction proof can be converted into a regular induction proof by modifying the proposition so that it holds for all values up to n. However, this makes the proof more complicated to read, so we just use strong induction instead. The term "strong" simply indicates that we are making a stronger assumption than in regular induction.

8.9 AN ANALOGY FOR UNDERSTANDING STRONG INDUCTION

Proofs by strong induction rely on the induction hypothesis being true for all smaller values of n. Continuing our tower analogy from earlier, this means that each level of the tower (except the bottom most ones) relies on direct support from levels one or more down from it.

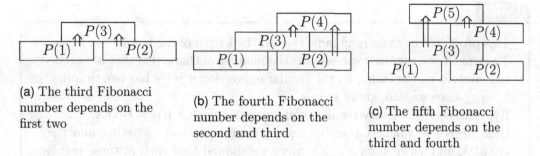

(a) The third Fibonacci number depends on the first two

(b) The fourth Fibonacci number depends on the second and third

(c) The fifth Fibonacci number depends on the third and fourth

Figure 8.2 The tower analogy for strong induction for the Fibonacci example

In an induction proof, we have to show the following two facts:

1. The bottom levels that are connected to the ground are firmly established. [Show $P(n)$ is true for all n that don't rely on $P(1), P(2), \ldots, P(n-1)$.]

2. If all levels $1, 2, 3, \ldots, k$ are firmly established, then the level $k+1$ will also be firmly established. [Show that if $P(1), P(2), P(3), \ldots, P(k)$ are all true then $P(k+1)$ is true.]

Sometimes we only need only some of $P(1), P(2), P(3), \ldots, P(k)$ to be true to prove that $P(k+1)$ is true. For example, in the Fibonacci proof, we only need the hypothesis to be true for the previous two values. The tower analogy for strong induction for the first few levels of the Fibonacci function is illustrated in Figure 8.2.

8.10 PROOFS WITH REGULAR EXPRESSIONS

We can use strong induction to prove the correctness of some of the procedures we designed for regular expressions in previous chapters. For example, in Chapter 6 we showed that any regular expression has a λ-NFA. We can prove this rigorously using strong induction.

Recall in Section 6.2 we saw a procedure for converting any regular expression into a λ-NFA. This involved designing λ-NFAs for the empty set, the set containing just the empty string, and for the set containing a one character string that will make up the base case of the following proof. In addition, there were procedures for designing λ-NFAs for the union and concatenation of any pair of expressions and the Kleene star of any expression; these are used in the inductive step of the following proof.

Theorem 8.5

The procedure outlined in Section 6.2 can be used to convert any regular expression to a λ-NFA.

Proof

This proof will be via strong induction on the length of the regular expression. The length of a regular expression is the number of characters that are needed to represent it. For example, the regular expression $a^*b + c$ has length 5 (a, *, b, $+$, c) since we will ignore spaces.

Hypothesis: $P(n)$: Every regular expression of length n has a λ-NFA.

Base step: The smallest possible regular expressions have length 1 and they are $\emptyset$, λ, and a (for each $a \in \Sigma$). Since we showed that each of these regular expressions has a λ-NFA in Figure 6.1, we have that $P(1)$ is true.

Inductive step: Fix some $k \geq 1$. We will assume that the hypothesis holds for all $n \leq k$ and show that $P(k + 1)$ is true.

Consider a regular expression E of size $k + 1 \geq 2$ (since $k \geq 1$). Since E has more than one symbol in it, by the definition of regular expressions, it must take one of the following forms:

1. $E = E_1 + E_2$ for some regular expressions E_1, E_2. Since the lengths of E_1 and E_2 are less than that of E, by the induction hypothesis there must exist λ-NFAs for both. We use the construction in Figure 6.2 to construct a λ-NFAs for E. More specifically, since the constructed λ-NFA has λ transitions to the start state of the λ-NFA for the languages of E_1 and E_2, the resulting λ-NFA accepts the union of these languages, as desired.

2. $E = E_1 \cdot E_2$ for some regular expressions E_1, E_2. Since the lengths of E_1 and E_2 are less than that of E, by the induction hypothesis there must exist λ-NFAs for both. We can then use the construction in Figure 6.3 to construct a λ-NFAs for E. More specifically, since the constructed λ-NFA has λ transitions from each of the accepting states of the λ-NFA for E_1 to the start state of E_2, the constructed λ-NFA will accept all strings in $E_1 \cdot E_2$, as desired.

3. $E = (E_1)^*$ for some regular expression E_1. Since the length of E_1 is less than that of E, by the induction hypothesis there must exist a λ-NFA for E_1. We can then use the construction in Figure 6.4 to construct a λ-NFAs for E. Since the constructed λ-NFA has λ transitions from each of the accepting states of the λ-NFA for E_1 to the starting state, the constructed λ-NFA will accept all strings in $(E_1)^*$, as desired.

Conclusion: We showed that $P(1)$ is true in the base case. Since we know that $P(1)$ is true, by the induction step this means that $P(2)$ is true. Since we know that $P(1)$ and $P(2)$ are true, this means that $P(3)$ is true. Continuing in this manner, we can show that $P(n)$ is true for all $n \geq 1$.

In our next example we will prove that the procedure for reversing a regular expression in Chapter 7 does indeed give the reverse of the original language.

Theorem 8.6

The procedure outlined in Section 7.5 can be used to convert any regular expression to a regular expression for the reverse of the language.

Proof

We will prove this via strong induction on the length of the regular expression.

Hypothesis: $P(n)$: Using the reverse procedure on any regular expression of length n gives the reverse of its language.

Base step: The shortest regular expressions are $\emptyset$, λ, and a (for each $a \in \Sigma$). The reverse of the language of each of these expressions is itself and Rule 1 of the procedure leaves them unchanged, so we have that $P(1)$ is true.

Inductive step: Fix some $k \geq 1$. We will assume that the hypothesis holds for all $n \leq k$ and show that $P(k+1)$ is true.

Consider a regular expression E of size $k+1$ which must be at least two since $k \geq 1$. Since E has more than one symbol in it, by the definition of regular expressions, it must take one of the following forms:

1. $E = E_1 + E_2$ for some regular expressions E_1, E_2. Since the lengths of E_1 and E_2 are less than that of E, by the induction hypothesis we know that the procedure correctly works for E_1 and E_2. As reversing the strings in each language and then taking their union is the same as taking the union and then reversing, we have that the expression $E_1^R + E_2^R$ will correctly give the reverse of the language of $E = E_1 + E_2$.

2. $E = E_1 \cdot E_2$ for some regular expressions E_1, E_2. Since the lengths of E_1 and E_2 are less than that of E, by the induction hypothesis we know that the procedure correctly works for E_1 and E_2. Then, the reverse of the language of $E = E_1 \cdot E_2$ must be strings that look like the reverse of a string from E_2 followed by the reverse of a string from E_1. This is precisely what is generated by $E_2^R \cdot E_1^R$.

3. $E = (E_1)^*$ for some regular expression E_1. Since the length of E_1 is less than that of E, by the induction hypothesis the expression for E_1^R must be the reverse of the language for E_1. Since the reverse of the language for $(E_1)^*$ consists of zero or more strings from E_1 concatenated together, the reverse of this language must be the reverse of zero or more strings from E_1 concatenated together, which is precisely what is generated by $(E_1^R)^*$.

Conclusion: We showed that $P(1)$ is true in the base case. Since we know that $P(1)$ is true, by the induction step this means that $P(2)$ is true. Since we know that $P(1)$ and $P(2)$ are true, this means that $P(3)$ is true. Continuing in this manner, we can show that $P(n)$ is true for all $n \geq 1$.

8.11   CORRECTNESS OF BINARY SEARCH

Another great application of strong induction is to more deeply understand the recursive binary search algorithm, shown in Algorithm 8.3. This algorithm assumes that the input list L is already sorted and gives an efficient way to search for the index of a target element by bisecting the region to be searched in half at every step. The recursive version of this algorithm also includes the left and right ends of the range of the indexes to be searched. When searching a list L indexed from 0 to $n-1$ for a target, the recursive function is called with parameters BinarySearch$(L, target, 0, n-1)$. The base cases of the recursion are when there are no more items to search (when the left index crosses the right index), in which case we return -1 to indicate that the search failed, and when the target is found at the middle of the active range of indices, in which case we return that middle index. Otherwise, the recursive procedure narrows the range to the appropriate half and continues the search.

Algorithm 8.3: BinarySearch$(L, target, left, right)$

> **input** : A sorted list L, an item being searched for $target$, the range of
> indexes of L to be searched $left, \ldots, right$
> **output:** An index where $target$ appears in L or -1 if $target$ is not in L
> 1: **if** $left > right$ **then**
> 2:   return -1
> 3: set $mid = (left + right)//2$ (rounding down)
> 4: **if** $target == L[mid]$ **then**
> 5:   return mid
> 6: **if** $target < L[mid]$ **then**
> 7:   return BinarySearch$(L, target, left, mid-1)$
> 8: **else**
> 9:   return BinarySearch$(L, target, mid+1, right)$

It can be notoriously hard to get the details of the indexes correct, which is why proving the correctness of the algorithm is so very important. For example, it is easy to mess up the indexes and end up in an infinite loop. We will show how to prove the correctness of this algorithm.

Unlike the earlier examples, it is not immediately clear here what should be the induction variable n. The size of the binary search problem isn't denoted by a single variable. If we think about how the problem is shrinking at each recursive call, we can see that the *range* of indices gets smaller each time. Therefore, we choose our induction variable to be the size of the range of indexes $n = right - left + 1$. For example, if the range is $left = 0$ and $right = 4$, then there are 5 indices in the range.

> ### Example 8.12
>
> Hypothesis: $P(n)$: The BinarySearch$(L, target, left, right)$ function call returns *index* such that $L[index] = target$ or -1 if $target$ is not in $L[left, \ldots, right]$ when $n = right - left + 1$.

The base case is when the list cannot be split any further, i.e., when $n = 0$.

(Base step) When $n = 0$, we have that $right - left + 1 = 0$ or $left = right + 1 > right$. Thus, the first condition (on Line 1) is true and the function returns -1, which is correct since $target$ is not in a sublist of size 0.

Example 8.14

(Inductive step) Fix some $k \geq 0$. Assume that the hypothesis holds for all $n \leq k$. We will show that $P(k+1)$ is true as well, that is the function returns the correct answer when $k + 1 = right - left + 1$.

When $k + 1 = right - left + 1$, we have that $k = right - left$, and we now have three cases for $mid = (left + right)//2$:

1. When $L[mid] = target$, the function correctly returns the value mid.

2. When $L[mid] > target$, the function makes a recursive call on Line 7 on a range of indexes of size $(mid - 1) - left + 1 = mid - left$, which must be smaller than $right - left + 1$ since $mid \leq right$, allowing us to use the inductive hypothesis on this recursive call. Now, there are two cases: either $target$ appears in $L[left, \ldots, right]$ or it does not. If $target$ appears in $L[left, \ldots, right]$, then we know that since L is sorted, $target$ must appear earlier than index mid. Thus, the recursive call on Line 7 must correctly return an index such that $L[index] = target$ by the induction hypothesis. If $target$ does not appear in $L[left, \ldots, right]$, then we know that the function will return -1 by the induction hypothesis. In both cases, it returns the correct answer.

3. When $L[mid] < target$, the function makes a recursive call on Line 9 on a range of indexes of size $right - (mid + 1) = right - mid - 1$, which must be smaller than $right - left + 1$ since $mid \geq left$, allowing us to use the inductive hypothesis on this recursive call. Again, there are two cases: either $target$ appears in $L[left, \ldots, right]$ or it does not. If $target$ appears in $L[left, \ldots, right]$, then we know that since L is sorted, $target$ must appear later than index mid. Thus, the recursive call on Line 9 must correctly return an index such that $L[index] = target$ by the induction hypothesis. If $target$ does not appear in $L[left, \ldots, right]$, then we know that the function will return -1 by the induction hypothesis. In both cases, it again returns the correct answer.

The wrapup step is similar to other strong induction proofs.

Example 8.15

Conclusion: We know that $P(0)$ is true from the base case. Since we know that $P(0)$ is true, by the induction step this means that $P(1)$ is true. Since we know that $P(0)$ and $P(1)$ are true, this means that $P(2)$ is true. Continuing in this manner, we can show that $P(n)$ is true for all $n \geq 0$.

Finally, we should connect the induction hypothesis we just proved with our original goal: to show correctness of the BinarySearch function.

Example 8.16

For any list L of length n with indices $0, \ldots, n - 1$, the function call BinarySearch$(L, target, 0, n-1)$ will return $index$ such that $L[index] = target$ or -1 if $target$ is not in $L[0, \ldots, n-1]$ since $P(n)$ is true.

8.12   CHAPTER SUMMARY AND KEY CONCEPTS

- **Induction** is a proof technique that allows us to prove many results in computer science, including showing that **recursive** algorithms are correct.

- An induction proof should be written with four parts:

  - A **hypothesis** $P(n)$ that states what you will prove.
  - A **base step** that shows $P(n)$ is true for small value(s) of n.
  - An **inductive step** that assumes $P(k)$ is true and shows that $P(k+1)$ is true.
  - A **conclusion step** that reminds you and the reader why the hypothesis $P(n)$ is true for all values of n.

- For any $a, b \in \mathbb{Z}, a \leq b$ and function f, the notation $\sum_{i=a}^{b} f(i)$ is a shortened way of expressing the sum $f(a) + f(a+1) + f(a+2) + \ldots + f(b)$.

- Induction is used for **analyzing run-time of algorithms** and for **counting the number of strings of a given length**.

- **Strong induction** makes a stronger assumption that the induction hypothesis is true for all values up to k before showing it to be true for $k+1$.

- Strong induction is used to show correctness of recursive problems in which the recursion operates on instances more than one smaller.

EXERCISES

Prove the following using induction:

8.1 Show that $\sum_{i=1}^{n}(3i-2) = \frac{n(3n-1)}{2}$ for all $n \geq 1$.

8.2 Show that $\sum_{i=1}^{n} i^2 = \frac{n(n+1)(2n+1)}{6}$ for all $n \geq 1$.

8.3 Show that $\sum_{i=1}^{n} i(i+2) = \frac{n(n+1)(2n+7)}{6}$ for all $n \geq 1$.

8.4 Show that $\sum_{i=1}^{n} i^3 = \left(\frac{n(n+1)}{2}\right)^2$ for all $n \geq 1$.

8.5 Show that $\sum_{i=1}^{n} \frac{1}{i(i+1)} = \frac{n}{n+1}$ for all $n \geq 1$.

8.6 Show that $3^0 + 3^1 + 3^2 + \ldots + 3^n = \frac{3^{n+1}-1}{2}$.

8.7 Show that for any $a, r \in \mathbb{R}, r \neq 1$, $a + ar + ar^2 + \ldots + ar^n = \frac{a(r^{n+1}-1)}{r-1}$.

8.8 Show that the sum of the first n odd natural numbers is equal to n^2 for all $n \geq 0$.

8.9 Write a recursive function to sum the first n positive integers and show that it is correct.

8.10 Write a recursive function to compute a^n for two positive integer parameters a, n and show that it is correct.

8.11 Write a recursive function to sum a list and show that it is correct.

8.12 Write a recursive function to find the minimum of a list and show that it is correct.

8.13 Write a recursive function to determine if a string is a palindrome and show that it is correct.

8.14 Write a recursive function to reverse a string and show that it is correct.

8.15 Write a recursive function to count the number of times a target item appears in a list and show that it is correct.

8.16 Write a recursive function that given a set as a list of unique elements returns the power set of the set (i.e., a list of lists) and show that it is correct. You may assume that all the elements in the list are unique.

8.17 Say that we have 3^n identically looking coins all of which weigh the same except one which is lighter than the others. Design a recursive procedure to find the lighter coin within n weighings on a balance scale. Prove this procedure correct using induction.

8.18 Show that $2^n > n$ for all $n \geq 1$.

8.19 Show that $2^n \geq n^2$ for all $n \geq 4$.

8.20 Find a value n_0 such that $2^n \geq n^3$ for all $n \geq n_0$ and then prove the result.

8.21 Show using induction that $\overline{S_1 \cup S_2 \cup \ldots \cup S_n} = \overline{S_1} \cap \overline{S_2} \cap \ldots \cap \overline{S_n}$ for all $n \geq 1$ and all sets $S_1, \ldots, S_n$. (Hint: Use De Morgan's laws in both the base step and the induction step.)

8.22 Show using induction that $\overline{S_1 \cap S_2 \cap \ldots \cap S_n} = \overline{S_1} \cup \overline{S_2} \cup \ldots \cup \overline{S_n}$ for all $n \geq 1$ and all sets $S_1, \ldots, S_n$. (Hint: Use De Morgan's laws in both the base step and the induction step.)

8.23 Prove that the size of the power set of a set S of size n is 2^n.

8.24 Show that it is possible to use postage stamps of values 2 and 3 to make any amount of postage $n \geq 3$.

8.25 Design your own analogy for induction similar to Section 8.3. Be sure to clearly state what are the hypothesis, base case, and inductive step in your analogy.

Prove the following using strong induction:

8.26 Let F_i be the ith Fibonacci number for each $i \geq 1$. Show that $\sum_{i=1}^{n} F_i = F_{n+2} - 1$.

8.27 Prove that the nth Fibonacci number F_n is equal to

$$\frac{\sqrt{5}}{5} \left(\left(\frac{1 + \sqrt{5}}{2} \right)^n - \left(\frac{1 - \sqrt{5}}{2} \right)^n \right).$$

8.28 Prove that for any $n \geq 1$ it is possible to tile a $2^n \times 2^n$ grid with L-shaped tiles (i.e., a 2×2 tile with one of the four corners missing) with just one corner square missing.

8.29 Show that it is possible to use postage stamps of values 3 and 5 to make any amount of postage $n \geq 8$.

8.30 Design your own analogy for strong induction. Be sure to clearly state what are the hypothesis, base case, and inductive step in your analogy

CHAPTER 9

Proving the Language of a DFA

9.1 WHY YOU SHOULD CARE

Now that you know how to create DFAs, you should next verify that the DFA you created does accept the language you think it does. This is analogous to showing that the algorithm you designed gives the correct answer on all possible inputs, something that every computer scientist should be doing.

9.2 A SIMPLE EXAMPLE

We will start with the simple DFA below (Figure 9.1):

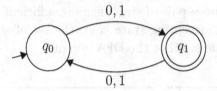

Figure 9.1 DFA for odd length strings over $\{0, 1\}$

It should be apparent that this DFA accepts strings of odd length. Why is this the case? Well, intuitively, the reason is that we flip-flop between the states q_0 and q_1 on any character, so that all even length strings end at q_0 and odd length strings end at q_1. We will formalize this idea next.

We first need to generalize the δ function to operate on strings rather than symbols. This will allow us say which state a string will end up at.

Definition 9.1

For any state of a DFA q_i and any string over the alphabet of the DFA s, we define $\hat{\delta}(q_i, s)$ to be the state that the DFA computation will end at if the string s is processed starting at state q_i.

DOI: 10.1201/9781003383284-9

More formally, we recursively define $\hat{\delta}(q_i, s)$ as follows:

$$\hat{\delta}(q_i, s) = \begin{cases} q_i, & \text{if } s = \lambda \\ \delta(\hat{\delta}(q_i, p), c), & \text{if } s = pc, \text{ where } p \in \Sigma^* \text{ and } c \in \Sigma. \end{cases}$$

Example 9.1

For the DFA in Figure 9.1, here are a few examples:

- $\hat{\delta}(q_0, \lambda) = q_0$

- $\hat{\delta}(q_0, 0) = q_1$

- $\hat{\delta}(q_0, 01) = q_0$

- $\hat{\delta}(q_0, 000) = q_1$

- $\hat{\delta}(q_0, 10101) = q_1$

- $\hat{\delta}(q_1, 101) = q_0$

Using this new notation, we can see that for the above DFA we have the following two criteria for the pair of states. For any string $s \in \Sigma^*$,

1. $\hat{\delta}(q_0, s) = q_0$ if and only if $|s|$ is even.

2. $\hat{\delta}(q_0, s) = q_1$ if and only if $|s|$ is odd.

Note that we always start from q_0 because this is the start state. It should be fairly clear why proving the above pair of statements is sufficient to prove the language of the DFA. Since the only accepting state is q_1, the set of strings that reach q_1—the odd length strings—is the set that the DFA accepts.

Theorem 9.1

The DFA in Figure 9.1 accepts the set of odd length strings over $\{0, 1\}$.

To prove this result, we need a more involved hypothesis, sometimes referred to as a **mutual induction** hypothesis. Notice that this hypothesis is over the length of the string n and thus must be shown true for every string of length n. It also has multiple parts, one per state of the DFA.

Example 9.2

(Hypothesis) For all $n \geq 0$, $P(n)$: for any string s of length n the following are both true:

1. $\hat{\delta}(q_0, s) = q_0$ if and only if $|s|$ is even.

2. $\hat{\delta}(q_0, s) = q_1$ if and only if $|s|$ is odd.

The base case of this induction hypothesis is more involved than usual because we have to prove if and only if results for both statements.

Example 9.3

(Base step) For the base case, we consider what happens when $n = 0$. The only string of length zero is λ. The first condition says that $\hat{\delta}(q_0, \lambda) = q_0$ if and only if 0 is even. This is a true statement since both are true and we know that $T \Leftrightarrow T$ is true. The second condition says that $\hat{\delta}(q_0, \lambda) = q_1$ if and only if 0 is odd. Interestingly, even though both these statements are false, the logical statement $F \Leftrightarrow F$ is true, so the second statement is correct as well.

Example 9.4

(Inductive step) We now assume that the proposition $P(n)$ is true for some $n \geq 0$. We will show that $P(n+1)$ is true as well.

Let $w = sc$ be a string over Σ of length $n + 1$ with $|s| = n$ and $|c| = 1$. Since $|s| = n$, we have that the induction hypothesis is true for s (by $P(n)$). We need to prove both statements of the proposition for w (of length $n + 1$) and do so in both directions (showing $P(n+1)$).

Proof of 1, if: We assume that w is an even length string. It then follows that s, which is one character shorter than w, is of odd length and from the induction hypothesis we must have that $\hat{\delta}(q_0, s) = q_1$. From the definition of δ, we know that $\hat{\delta}(q_0, w) = \delta(\hat{\delta}(q_0, s), c) = \delta(q_1, c) = q_0$.

Proof of 1, only if: We assume that $\hat{\delta}(q_0, w) = q_0$. We have two cases: when the last character c is a 0 or a 1. In either case, if $c = 0$ or $c = 1$, the only state that we could have come from to arrive at q_0 was q_1, i.e, $\hat{\delta}(q_0, s) = q_1$. Since s is of length n, from the induction hypothesis we know that s is of odd length. Since w is one character longer than s, it follows that w has an even length.

The proofs for the second part of the hypothesis are very similar and you can closely follow much of the language. Pay close attention to what has changed from the proof of the first part and be sure to understand why the changes are needed.

Example 9.5

Proof of 2, if: We assume that w is an odd length string. It then follows that s, which is one character shorter than w, is of even length and from the induction hypothesis we have $\hat{\delta}(q_0, s) = q_0$. From the definition of δ, we know that $\hat{\delta}(q_0, w) = \delta(\hat{\delta}(q_0, s), c) = \delta(q_0, c) = q_1$.

Proof of 2, only if: We assume that $\hat{\delta}(q_0, w) = q_1$. We have two cases: when the last character c is a 0 or a 1. In either case, if $c = 0$ or $c = 1$, the only state that we could have come from to arrive at q_1 was q_0, i.e, $\hat{\delta}(q_0, s) = q_0$. Since s is of length n, from the induction hypothesis we know that s is of even length. Since w is one character longer than s, it follows that w has an odd length.

The induction wrapup is the same as the previous proofs by induction that you have seen before. However, there is an additional step where you should tell your reader why the hypothesis proves that the DFA does indeed accept the desired language. Specifically, you should tell the reader the language of the DFA, which is all the strings that can reach an accepting state.

Example 9.6

(Wrapup) We know that $P(0)$ is true from the base case. Since $P(0)$ is true and we showed that $P(0)$ implies $P(1)$ in the inductive step (when $k = 0$), this means that $P(1)$ is true. Since $P(1)$ is true and we showed that $P(1)$ implies $P(2)$ in the inductive step (when $k = 1$), this means that $P(2)$ is true. Continuing in this manner, we can show that $P(n)$ is true for all $n \geq 0$.

Note that the only accepting state in the DFA is q_1 and hence the only strings it accepts are the ones that reach this state. It follows then that the DFA accepts precisely the strings of odd length from Part 2 of $P(n)$.

9.3   A MORE INVOLVED EXAMPLE

The above example was simple in that the DFA had the same transition for both symbols of the alphabet from both states. Let us look at a slightly more complex example in which this is not the case.

Take a look at the DFA in Figure 9.2 and think about what language it accepts.

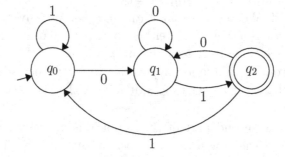

Figure 9.2 DFA for all strings over $\{0, 1\}$ that end in 01

After giving it some thought, hopefully you came to the conclusion that this DFA accepts precisely the set of strings over $\Sigma = \{0, 1\}$ that end in 01. Intuitively, the strings that reach state q_1 must be precisely the ones that end in 0 and thus the strings that reach state q_2 must be precisely the ones that end in 01. By the process of elimination, this means that the strings that reach state q_0 are the ones that neither end in 0 nor 01. We use this to once again prove the result as above.

Theorem 9.2

The DFA in Figure 9.2 accepts the set of strings over $\{0, 1\}$ that end in 01.

Example 9.7

We will prove this assertion using mutual induction.

Hypothesis: For all $n \geq 0$, $P(n)$: for any string s of length n these are all true:

1. $\hat{\delta}(q_0, s) = q_0$ if and only if s does not end in 0 or 01.

2. $\hat{\delta}(q_0, s) = q_1$ if and only if s ends in 0.

3. $\hat{\delta}(q_0, s) = q_2$ if and only if s ends in 01.

Example 9.8

Base step: For the base case, we consider what happens when $n = 0$. The only string of length zero is λ. The first condition says that $\hat{\delta}(q_0, \lambda) = q_0$ if and only if s does not end in 0 or 01. This is a true statement since both are true and $T \Leftrightarrow T$ is true. The second condition says that $\hat{\delta}(q_0, \lambda) = q_1$ if and only if s ends in 0, which is true since $F \Leftrightarrow F$ is true. The third condition says that $\hat{\delta}(q_0, \lambda) = q_2$ if and only if s ends in 01, which is true since $F \Leftrightarrow F$ is true.

Example 9.9

Inductive step: We now assume that the proposition $P(n)$ is true for some $n \geq 0$. We will show that $P(n + 1)$ is true as well.

Let $w = sc$ be a string over Σ of length $n + 1$ with $|s| = n$ and $|c| = 1$. We assume the hypothesis is true for s (assuming $P(n)$) and need to prove all three parts of the proposition in both directions for w (show $P(n + 1)$).

Proof of 1, if: We assume that w does not end in 0 or 01. It follows then that $c = 1$ and s is either the empty string or s ends in 1. If s is the empty string, then $w = sc = 1$ and so $\hat{\delta}(q_0, w) = q_0$. If s ends in 1, then by the induction hypothesis $\hat{\delta}(q_0, s)$ must be q_0 or q_2 and so $\hat{\delta}(q_0, w) = q_0$ since $\delta(q_0, 1)$ and $\delta(q_2, 1)$ are both q_0.

Proof of 1, only if: We assume that $\hat{\delta}(q_0, s) = q_0$. It is not possible for c to be 0 since there are no transitions entering q_0 on a 0. If $c = 1$, then $\hat{\delta}(q_0, s)$ is q_0 or q_2. By the induction hypothesis, s does not end in 0, so w cannot end in 01.

Proof of 2, if: We assume that w ends in 0. Since all three states transition to q_1 on a 0, this means that $\hat{\delta}(q_0, w) = q_1$.

Proof of 2, only if: Assume that $\hat{\delta}(q_0, w) = q_1$. Then c cannot be 1 as there are no δ transitions into q_1 on a 1. Thus w must end in a 0.

Proof of 3, if: Assume that w ends in 01. Then, s ends in 0 and so by the induction hypothesis, $\hat{\delta}(q_0, s) = q_1$. Since $\delta(q_1, 1) = q_2$, we have $\hat{\delta}(q_0, w) = q_2$.

Proof of 3, only if: Assume $\hat{\delta}(q_0, w) = q_2$. The only transition into q_2 is from q_1 on a 1, so c must be 1. Also since $\hat{\delta}(q_0, s) = q_1$, by the induction hypothesis s must end in a 0. Thus, w ends in 01.

We wrap the proof up similar to before.

(Wrapup) We know that $P(0)$ is true from the base case. Since we know that $P(0)$ is true and we showed that $P(0)$ implies $P(1)$ in the inductive step (when $k = 0$), this means that $P(1)$ is true. Since we know that $P(1)$ is true and we showed that $P(1)$ implies $P(2)$ in the inductive step (when $k = 1$), this means that $P(2)$ is true. Continuing in this manner, we can show that $P(n)$ is true for all $n \geq 0$.

The last step of our proof is to note that the only accepting state in the DFA is q_2 and hence the only strings it accepts are the ones that end in 01, from Part 3 of $P(n)$.

9.4   AN EXAMPLE WITH SINK STATES

Here is another example with sink states.

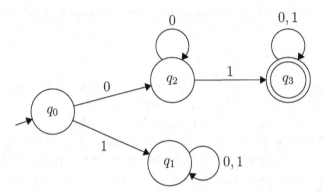

Figure 9.3 DFA for all strings over $\{0, 1\}$ that start with 0 and contain a 1

Before reading on, try to determine the set of strings that can reach the states q_0, q_1, q_2, and q_3 in Figure 9.3. Some examples of strings in each set are given below to help you get started.

Following the $\hat{\delta}$ notation from earlier, here is where several strings reach:

- $\hat{\delta}(q_0, \lambda) = q_0$

- $\hat{\delta}(q_0, 0) = \hat{\delta}(q_0, 00) = \hat{\delta}(q_0, 000) = \hat{\delta}(q_0, 0000) = q_2$

- $\hat{\delta}(q_0, 1) = \hat{\delta}(q_0, 10) = \hat{\delta}(q_0, 11) = q_1$

- $\hat{\delta}(q_0, 01) = \hat{\delta}(q_0, 001) = \hat{\delta}(q_0, 010) = \hat{\delta}(q_0, 001011) = q_3$

Here are the sets for each state:

- The start state q_0 has no transitions entering it, so the only string that ends at q_0 is λ. Thus, the set in this case is $\{\lambda\}$.

- The state q_1 is a sink state for strings that start with a 1. The set is then $\{1s : s \in \{0,1\}^*\}$.

- The state q_2 can only be reached by strings that have one or more 0s and no 1s. The set is thus $\{0^n : n \geq 1\}$.

- Lastly, the state q_3 has all the remaining strings, namely the ones that start with a 0 and contain a 1. This set is $\{0s1t : s, t \in \{0,1\}^*\}$.

Theorem 9.3

The DFA in Figure 9.3 accepts the set of strings over $\{0,1\}$ that start with 0 and contain a 1.

Example 9.12

We will prove this assertion using mutual induction with the following hypothesis.

(Hypothesis) For all $n \geq 0$, $P(n)$: for any string s of length n these are all true:

1. $\hat{\delta}(q_0, s) = q_0$ if and only if $s = \lambda$.

2. $\hat{\delta}(q_0, s) = q_1$ if and only if s starts with a 1.

3. $\hat{\delta}(q_0, s) = q_2$ if and only if $s \in \{0^n : n \geq 1\}$.

4. $\hat{\delta}(q_0, s) = q_3$ if and only if s starts with a 0 and contains a 1.

Example 9.13

(Base step) For the base case, we consider what happens when $n = 0$. The only string of length zero is λ. The first condition says that $\hat{\delta}(q_0, \lambda) = q_0$ if and only if $s = \lambda$. This is a true statement since both are true and we know that $T \Leftrightarrow T$ is true.

The second condition says that $\hat{\delta}(q_0, \lambda) = q_1$ if and only if s starts with a 1, which is true since both are false and $F \Leftrightarrow F$ is true.

The third condition says that $\hat{\delta}(q_0, \lambda) = q_2$ if and only if $s \in \{0^n : n \geq 1\}$, which is true since both are false and $F \Leftrightarrow F$ is true.

The fourth condition says that $\hat{\delta}(q_0, \lambda) = q_3$ if and only if s starts with a 0 and contains a 1, which is true since both are false and $F \Leftrightarrow F$ is true.

Example 9.14

(Inductive step) We now assume that the proposition $P(n)$ is true for some $n \geq 0$. We will show that $P(n + 1)$ is true as well.

Let $w = sc$ be a string over Σ of length $n + 1$ with $|s| = n$ and $|c| = 1$. We assume the hypothesis is true for s (assuming $P(n)$) and need to prove all four parts of the proposition in both directions for w (show $P(n + 1)$).

<u>Proof of 1</u>: Since $|w| = n + 1 > 0$, it is impossible for w to be equal to λ. Moreover, we can confirm from the diagram of the DFA that no string of length greater than zero can end at state q_0.

<u>Proof of 2, if</u>: We assume that w starts with a 1. Then, s must also start with a 1 as $w = sc$. Thus, $\hat{\delta}(q_0, s) = q_1$ by the induction hypothesis. Since the only transitions from q_1 are to itself and $w = sc$, $\hat{\delta}(q_0, w) = q_1$.

<u>Proof of 2, only if</u>: Assume that $\hat{\delta}(q_0, w) = q_1$. Then by the definition of δ, either $\hat{\delta}(q_0, s) = q_0$ or $\hat{\delta}(q_0, s) = q_1$. If $\hat{\delta}(q_0, s) = q_0$, then by the induction hypothesis $s = \lambda$ and thus c must be a 1, which means that w starts with a 1. If $\hat{\delta}(q_0, s) = q_1$, then by the induction hypothesis s starts with a 1 and thus $w = sc$ also starts with a 1.

<u>Proof of 3, if</u>: Assume that $w \in \{0^n : n \geq 1\}$. Then since $w = sc$, $c = 0$ and $s \in \{0^n : n \geq 0\}$. If $s = \lambda$, then by the induction hypothesis $\hat{\delta}(q_0, s) = q_0$ and thus $\hat{\delta}(q_0, w) = q_2$ (as $c = 0$). If $s \in \{0^n : n \geq 1\}$, then by the induction hypothesis, $\hat{\delta}(q_0, s) = q_2$ and since $c = 0$, $\hat{\delta}(q_0, s) = q_2$.

<u>Proof of 3, only if</u>: Assume $\hat{\delta}(q_0, w) = q_2$. The only transition into q_2 is from q_0 or q_2 on a 0. In the former case, by the induction hypothesis we must have $s = \lambda$ and $c = 0$, so $w = 0 \in \{0^n : n \geq 1\}$. In the latter case, since $\hat{\delta}(q_0, s) = q_2$ and $c = 0$, by the induction hypothesis $s \in \{0^n : n \geq 1\}$, so $w = sc = s0 \in \{0^n : n \geq 1\}$.

<u>Proof of 4, if</u>: Assume that w starts with a 0 and contains a 1. It must follow that s also starts with a 0. By the induction hypothesis, either $\hat{\delta}(q_0, s) = q_2$ or $\hat{\delta}(q_0, s) = q_3$ since strings that start with 0 cannot end at q_0 or q_1. If $\hat{\delta}(q_0, s) = q_2$, then by the induction hypothesis s has no 1s and so $c = 1$ (since w contains a 1) and thus, $\hat{\delta}(q_0, w) = q_3$. If $\hat{\delta}(q_0, s) = q_3$, then as the only transitions out of q_3 are to itself, this means that $\hat{\delta}(q_0, w) = q_3$.

<u>Proof of 4, only if</u>: Assume $\hat{\delta}(q_0, w) = q_2$. By the definition of δ, we can see that either $\hat{\delta}(q_0, s) = q_2$ or $\hat{\delta}(q_0, s) = q_3$. If $\hat{\delta}(q_0, s) = q_2$, then $c = 1$ and by the induction hypothesis $s \in \{0^n : n \geq 1\}$, so w starts with a 0 and contains a 1. If $\hat{\delta}(q_0, s) = q_3$, then by the induction hypothesis we know that s starts with a 0 and contains a 1, so the same must be true of $w = sc$.

Finally, we wrap up the induction step and explain to the reader how the induction hypothesis tells us the language of the DFA.

Example 9.15

(Wrapup) We know that $P(0)$ is true from the base case. Since we know that $P(0)$ is true and we showed that $P(0)$ implies $P(1)$ in the inductive step (when $k = 0$), this means that $P(1)$ is true. Since we know that $P(1)$ is true and we showed that $P(1)$ implies $P(2)$ in the inductive step (when $k = 1$), this means that $P(2)$ is true. Continuing in this manner, we can show that $P(n)$ is true for all $n \geq 0$.

The last step of our proof is to note that the only accepting state in the DFA is q_3 and hence the only strings it accepts are the ones that start with 0 and contain a 1, from Part 4 of $P(n)$.

9.5   CHAPTER SUMMARY AND KEY CONCEPTS

- We can prove that a DFA accepts a given language using **mutual induction.**

- The function $\hat{\delta}$ extends the δ function of a DFA to apply to strings.

- The mutual induction **hypothesis** defines precisely what set of strings ends at each state.

- The mutual induction **base case** verifies the hypothesis for the empty string.

- The mutual induction **inductive step** proves that strings of length $n + 1$ end at the correct state assuming that the strings of length n do so.

- The mutual induction **conclusion** both concludes the induction proof and demonstrates the language of the DFA by examining the set of strings that reach the accepting state(s).

EXERCISES

Draw DFAs with as few states as possible for each of the following languages and then prove that they accept that language.

9.1 $\{w \in \{0,1\}^* : w \text{ starts with a } 0\}$

9.2 $\{w \in \{0,1\}^* : w \text{ ends with a } 1\}$

9.3 $\{w \in \{0,1\}^* : w \text{ has 01 as a substring}\}$

9.4 $\{w \in \{0,1\}^* : w \text{ does not have 01 as a substring}\}$

9.5 $\{w \in \{0,1\}^* : w \text{ has an even number of 0s}\}$

9.6 $\{w \in \{0,1\}^* : w \text{ has an even number of 0s and an odd number of 1s}\}$

9.7 $\{w \in \{0,1\}^* : w \text{ has an even number of 0s and an even number of 1s}\}$

9.8 $\{w \in \{0,1\}^* : w \text{ has an odd number of 0s and an even number of 1s}\}$

9.9 $\{w \in \{0,1\}^* : w$ has an odd number of 0s and an odd number of 1s$\}$

9.10 $\{w \in \{0,1\}^* : w$ starts with 00$\}$

9.11 $\{w \in \{0,1\}^* : w$ starts and ends with a 0$\}$

9.12 $\{w \in \{0,1\}^* : w$ has exactly one 0$\}$

9.13 $\{w \in \{0,1\}^* :$ the number of 0s in w is divisible by 3$\}$

9.14 $\{w \in \{0,1\}^* :$ the second character in w is 0$\}$

9.15 $\{w \in \{0,1\}^* :$ the second last character in w is 0$\}$

CHAPTER 10

Proof by Contradiction

10.1 WHY YOU SHOULD CARE

Proof by contradiction is a powerful technique that is often the best way to prove many results in computer science. It can be an easier way to prove some results than, say, direct proof because it allows the prover to assume more facts to be used in the proof.

10.2 OVERVIEW OF THE TECHNIQUE

Proof by contradiction works by assuming the opposite (negation) of what you are trying to prove and then arriving at a contradictory statement (usually that a statement and its negation are both true). In other words, the structure of proving some statement S looks like the following:

1. Assume for a contradiction that the negation of S is true.

2. Make arguments based on this assumption.

3. Show that you arrive at a contradiction.

This technique gives you a starting point for your proof—assuming the negation of the statement you are trying to prove. The tricky part with it is figuring out how to get to the final contradiction.

10.3 WHY YOU CAN'T WRITE $\sqrt{2}$ AS AN INTEGER FRACTION

The technique of proof by contradiction dates back at least to the ancient Greek who used it to show that $\sqrt{2}$ is irrational.

Definition 10.1
A real number is called **rational** if it can be written as p/q, where $p, q \in \mathbb{Z}$ and $q \neq 0$.

Definition 10.2

A real number is called **irrational** if it cannot be written as p/q, where $p, q \in \mathbb{Z}$ and $q \neq 0$.

The Pythagoreans used proof by contradiction to show that $\sqrt{2}$ is irrational. An interesting side effect of this is that it is impossible to precisely store the value of $\sqrt{2}$ as a `float`, `double`, or even using a rational number class since we would need infinite precision to capture all the digits in the number. We see why next.

Theorem 10.1

The value $\sqrt{2}$ is irrational.

Proof

Assume for a contradiction that $\sqrt{2}$ is not irrational. Then, there exists $p, q \in \mathbb{Z}$ ($q \neq 0$) such that $p/q = \sqrt{2}$. We will assume that p and q are reduced to the lowest terms, that is they have no common factors. If p and q did have any common factors, we can simply divide both by this common factor and use the resulting numbers as our new p and q.
Since we have $p/q = \sqrt{2}$, we can square both sides and rearrange to get $p^2 = 2q^2$. This implies that p^2 is even as the right side of this equation has a factor of 2 in it. This tells us that p must be even since the square of an odd number is always an odd number. Thus, we can write p as $p = 2r$, where $r \in \mathbb{Z}$. Substituting this into the equation $p^2 = 2q^2$, we get $(2r)^2 = 2q^2$ or $4r^2 = 2q^2$ or $2r^2 = q^2$. As before, we can show that q must be an even number. This leads us to a contradiction as we have shown that both p and q are divisible by 2, even though we had assumed that they had no common factors. Thus, $\sqrt{2}$ cannot be written in the form p/q ($p, q \in \mathbb{Z}$) and must be irrational.

In the above proof we use the fact that the square of an odd number is always an odd number. This follows from the fundamental theorem of arithmetic.

Definition 10.3

The **fundamental theorem of arithmetic** states that every integer greater than one can be written as the product of prime numbers in a unique manner (disregarding other orderings of the primes).

Example 10.1

The number 700 can be written as $700 = 2^2 \cdot 5^2 \cdot 7$. Other than rearranging the factors, this is the only way to factor 700 into primes.

We can use this theorem to see that since an odd number has no factor of 2 in its prime factorization, then neither will its square.

10.4 WILL WE RUN OUT OF PRIME NUMBERS?

A number of algorithms, such as several for encrypting data, rely on the existence of large prime numbers. It would be problematic if there were only a finite number of prime numbers. This next proof shows that this is not the case.

Theorem 10.2

There are infinitely many prime numbers.

Proof

Assume for a contradiction that there are finitely many prime numbers. Let us call this finite number k. Then we can write out the list of prime numbers in ascending order as $p_1 = 2, p_2 = 3, p_3 = 5, \ldots, p_k$.

Consider the number $M = p_1 p_2 p_3 \ldots p_k + 1$. This number is not divisible by any of the above primes p_i $(1 \leq i \leq k)$ since dividing M by p_i always gives a remainder of one, rather than a remainder of zero. This means that M is not divisible by any of the known primes. This can only happen if M has some unknown prime factors or is a prime number itself, but in both cases this contradicts the fact that we could list out all the primes. Hence, there cannot be a finite number of prime numbers.

10.5 THE MINDBENDING NUMBER OF LANGUAGES

The next result we will see is that the number of possible languages is so large that we cannot even list them all.

Theorem 10.3

It is not possible to list all possible languages over $\{0,1\}$ in a list of the form $L_1, L_2, L_3, \ldots$.

Proof

Assume for a contradiction that all possible languages over $\{0,1\}$ can be listed as $L_1, L_2, L_3, \ldots$. Let us arrange the strings over $\{0,1\}$ in shortlex order and call them $s_1 = \lambda, s_2 = 0, s_3 = 1, s_4 = 00, \ldots$.

We will create a new language L' that is not in the list $L_1, L_2, L_3, \ldots$ as follows: For each string s_i, if $s_i \notin L_i$ we add it to L', otherwise we omit it.

We can now show that this language L' is not in the list. Say it appeared in the list as L_j for some $j \geq 1$. By the above definition, we know that s_j is in L' if and only if s_j is not in L_j. This means that L' cannot be the same as L_j, a contradiction.

Thus, the set of languages over $\{0,1\}$ is so large that we cannot even list all of them. This is called uncountably infinite.

This proof is an adaptation of Cantor's diagonalization technique that showed that the real numbers cannot be listed either, which means that there are different sizes of infinity. This technique will be important for us in Chapter 14 when we show that there are languages that cannot be accepted by any computer.

10.6 CHAPTER SUMMARY AND KEY CONCEPTS

- The **proof by contradiction** technique is used to show a statement is true by starting with the negation of the statement and ending in an impossible situation (contradiction).

- The square root of two cannot be written as a fraction of integers.

- There are infinitely many prime numbers.

- The number of possible languages is so large that we cannot even make an infinite list of them.

EXERCISES

10.1 Prove that $\sqrt{3}$ is irrational.

10.2 Explain why the proof used to show that $\sqrt{2}$ is irrational cannot be adapted to show that $\sqrt{4}$ is irrational.

10.3 Prove that for any prime number p, $\sqrt{p}$ is irrational.

10.4 Prove that the cube root of two ($\sqrt[3]{2}$) is irrational.

10.5 Prove that if a, b are integers, then $a^2 - 4b \neq 2$.

10.6 Prove that if a, b, c are integers and $a^2 + b^2 = c^2$, then a or b is even.

10.7 Prove that $\log_2 3$ is irrational.

10.8 Prove that $\log_{10} 8$ is irrational.

10.9 Prove that any number $x \neq 0$ is irrational if and only if $1/x$ is irrational.

10.10 Prove that the sum of a rational number and an irrational number is an irrational number.

10.11 Prove that there is an irrational number x such that $x^{\sqrt{2}}$ is rational.

10.12 Define a majority item in a list to be an item that makes up **more** than half the list. Prove that there can be at most one majority item in the list.

10.13 Define a majority item in a list to be an item that makes up **more** than half the list. Let L be an even length list with a majority item. Prove that any way that L is split into two equal halves, one of the halves must have the same majority item as a majority item of the entire list.

10.14 Prove that the average of two numbers is at least as large as at least one of the two numbers.

10.15 Prove that the average of n numbers is at most as large as at least one of the numbers.

10.16 Prove that all positive integers can be written as a sum of unique, integer powers of 2. (Hint: You will need strong induction and contradiction for this proof.)

CHAPTER 11

Pumping Lemma for Regular Languages

11.1 WHY YOU SHOULD CARE

Now that you have learned about proof by contradiction, you will see how to apply it to show that certain languages are not regular—no matter how hard you try, you cannot create a DFA for them. This is our first taste of what is not computable with (a specific model of) computers. It also means that there are fairly simple problems for which we cannot design a DFA and thus DFAs are not the model of computation that we want to use to represent all computers.

11.2 THE PIGEONHOLE PRINCIPLE

Before we can see the proof technique for showing a language is not regular, we need to understand a mathematical idea called the pigeonhole principle. This idea sounds obvious but turns out to be a very powerful tool for showing surprising results.

> **Definition 11.1**
>
> The pigeonhole principle states that if more than n objects are assigned to n containers, then some container must have more than one of the objects.

Here are some interesting examples:

> **Example 11.1**
>
> Any collection of five socks of four different colors must have a matched pair.

> **Example 11.2**
>
> Any group of 13 people has two that have the same birth month. Any group of 367 people has two that have the same birthday (allowing for leap years).

DOI: 10.1201/9781003383284-11

Example 11.3

Suppose everyone has fewer than 200,000 hairs on their head. Then any group of 200,001 people contains at least one pair with the same number of hairs on their head.

We can use the pigeonhole principle to show that a language is not regular as follows.

Theorem 11.1

The language $L = \{0^n 1^n : n \geq 0\}$ is not regular.

Proof

Assume for a contradiction that the language $L = \{0^n 1^n : n \geq 0\}$ is regular. This means that there exists a DFA for the language L with, say, $k > 0$ states. Consider the strings $\lambda, 0, 00, 000, \ldots, 0^k$. Since there are $k+1$ strings (including the empty string of length 0), two of these strings must end at the same state by the pigeonhole principle. Let us call these strings 0^i and 0^j, where $i \neq j$. The DFA must end at the same ending state for the strings $0^i 1^i$ and $0^j 1^i$ since both strings arrive at the same state after 0^i and 0^j and then process the same string (1^i). Now, the string $0^i 1^i$ should reach an accepting state since it is in the language L but the string $0^j 1^i$ should not since $i \neq j$. Since the same state cannot be accepting and non-accepting, this is a contradiction. Thus, L is not regular.

This proof very neatly shows why the language is not regular. However, it is not easy to generalize to other languages that are not regular. A more general method is to use the pumping lemma.

Theorem 11.2: Pumping Lemma

If L is a regular language, then there exists some $n > 0$ such that for all strings $w \in L$ such that $|w| \geq n$, we can break w into $w = xyz$ such that (i) $y \neq \lambda$, (ii) $|xy| \leq n$, and (iii) for all $k \geq 0$, $xy^k z \in L$.

At a high level, the pumping lemma says that for any regular language, if we take any sufficiently long string in the language, we can break it into three parts (with some conditions on the lengths of these three parts) in such a way that the string can be pumped—repeating the middle of the three parts any number of times (including zero) must always give us back a string in the language. This pumping condition is what gives the lemma its name.

If we can show that the pumping lemma does not apply to a language L, then this means (by contradiction) that L is not regular. Before we see an example of this, we will prove the pumping lemma.

Proof

Let L be a regular language. Then L must have a DFA with (say) n states. For any string $w \in L$ such that $|w| = m \geq n$, let us denote the characters of the string as $w_1, w_2, \ldots, w_m$, that is $w = w_1 w_2 \ldots w_m$, where each $w_i \in \Sigma$ $(1 \leq i \leq m)$.

We consider the prefixes of w of length at most n. Since there are $n+1$ of them (including the prefix of length zero), by the pigeonhole principle we know that two of these prefixes end at the same state. Let us denote the length of these prefixes by i and j $(0 \leq i < j \leq n)$.

We now set $x = w_1 \ldots w_i$, $y = w_{i+1} \ldots w_j$ and $z = w_{j+1} \ldots w_m$ and show that all three criteria of the pumping lemma hold true for this choice of x, y, and z.

First, $y \neq \lambda$ because we said that $i < j$ and so the length of y (which is $j - i$) is greater than zero.

Second, note that we said that $j \leq n$ so that means that $|xy|$ must be no larger than n.

Third, we will show that the string can be pumped any number of times— this is how the pumping lemma got its name. Recall that we said that the strings $w_1 \ldots w_i$ and $w_1 \ldots w_j$ both end at the same state in the DFA, call it q. This means that the DFA transitions from q back to itself on the string $y = w_{i+1} \ldots w_j$. This is illustrated below:

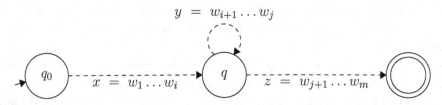

We will divide the rest of the proof up into two cases:

Case 1: $(k = 0)$ If we pump the string $k = 0$ times, then we have the string $xy^0 z = xz = w_1 \ldots w_i w_{j+1} \ldots w_m$. We know that the DFA accepts when it starts at state q and encounters $w_{j+1} \ldots w_m$ because it accepts $w = w_1 \ldots w_j w_{j+1} \ldots w_m$. Thus, xz must also be accepted.

Case 2: $(k > 1)$ For any $k > 1$, the pumped string is $xy^k z$ and we know that after $x = w_1 \ldots w_i$ we get to state q. After each instance of $y = w_{i+1} \ldots w_j$, we know that we return to the state q. Finally, as before, from state q on the string $z = w_{j+1} \ldots w_m$ we get to an accepting state because we know that $w = w_1 \ldots w_j w_{j+1} \ldots w_m$ is accepted. Thus, $xy^k z$ must also be accepted.

This concludes the proof for the third condition. Since we have shown all three conditions are true, the pumping lemma must be true.

A question you may have is what happens if a language does not have arbitrarily long strings. For example, a simple language like $\{aa, ab\}$ should be regular (it has a regular expression $aa + ab$) but does not have strings larger than n for any $n > 2$.

This is an example of a vacuously true statement. For this language, we can pick $n = 3$ and note that since there are no strings longer than length 2 for this language, the pumping lemma conditions apply for all these strings (all zero of them!).

11.3   APPLYING THE PUMPING LEMMA

The general form of the proof to show that a language is not regular is to assume for a contradiction that it is regular and then show that the pumping lemma does not apply to it for a contradiction. To show why the pumping lemma doesn't apply, we have to show its negation.

The pumping lemma takes the following form for a regular language L:

$$(\exists n > 0)$$
$$(\forall \text{ strings } w \text{ in } L \text{ of length at least } n)$$
$$(\exists \text{ a way to break up the string into } w = xyz \text{ with } |y| > 0 \text{ and } |xy| \leq n)$$
$$(\forall k \geq 0)(xy^k z \in L).$$

To show the negation, we need to negate the above statement. We can do so using the quantifier negation that you learned about in Chapter 3. This gives the following:

$$(\forall n > 0)$$
$$(\exists \text{ a string } w \text{ in } L \text{ of length at least } n)$$
$$(\forall \text{ ways to break up the string into } w = xyz \text{ with } |y| > 0 \text{ and } |xy| \leq n)$$
$$(\exists k \geq 0)(xy^k z \notin L).$$

We are now ready to see how we can use the pumping lemma to show that the same language from earlier is not regular.

Theorem 11.3

The language $L = \{0^n 1^n : n \geq 0\}$ is not regular.

Proof

Let us assume for a contradiction that L is regular. Then the pumping lemma applies to it for some $n > 0$. We pick the string $w = 0^n 1^n$ and note that $w \in L$ and that $|w| = 2n \geq n$.

We now consider all possible ways to break up w into $w = xyz$ such that $|y| > 0$ and $|xy| \leq n$. Since the first n symbols in w are all 0s, the strings x and y can only have 0s in them.

Let us say that $x = 0^i$ and $y = 0^j$, where $i \geq 0$ and $j \geq 1$. This means that the remaining symbols in w go into $z = 0^{n-i-j} 1^n$.

We now have to pick a $k \geq 0$ such that $xy^k z$ is not in L. Selecting $k = 0$, we get that $xy^k z = xz = 0^i 0^{n-i-j} 1^n = 0^{n-j} 1^n$, which is not in L since $j > 0$ so $n - j \neq n$. This contradicts the pumping lemma. Thus, the language L is not regular.

When using the pumping lemma, there are three places in which you have to use your insight into how these proofs work to get a correct argument. These are (i) correctly selecting an appropriate w, (ii) correctly splitting w into x, y, and z in all possible ways, and (iii) correctly selecting an appropriate k for each case. The next few examples will give some insights into how to solve each of these problems.

11.4   SELECTING THE STRING FROM THE LANGUAGE

This example will help us see a way to choose the string in the language.

Theorem 11.4

The language $L = \{s \in \{a, b\}^* : s$ has an equal number of as and bs$\}$ is not regular.

Proof

Let us assume for a contradiction that L is regular. Then the pumping lemma applies to it for some $n > 0$. Now, there are many ways to select the string w. To make our cases as simple as possible, similarly to the last example, we can pick the string $w = a^n b^n$ and note that $w \in L$ since it has equal as and bs and that $|w| = 2n \geq n$.

We now consider all possible ways to break up w into $w = xyz$ such that $|y| > 0$ and $|xy| \leq n$. Since the first n symbols in w are all as, the strings x and y can only have as in them.

Let us say that $x = a^i$ and $y = a^j$, where $i \geq 0$ and $j \geq 1$. This means that the remaining symbols in w go into $z = a^{n-i-j}b^n$.

We now pick a $k \geq 0$ such that $xy^k z$ is not in L. Selecting $k = 0$, we get that $xy^k z = xz = a^i a^{n-i-j} b^n = a^{n-j} b^n$, which is not in L since $j > 0$ so there are fewer as than bs. This contradicts the pumping lemma. Thus, L is not regular.

Notice that for this language we had any number of strings that we could have chosen. The string we chose was done so deliberately so as to make the cases for how to split w into xyz as simple as possible since x and y had to consist entirely of as. This may not always be the case as we'll see in the next example.

11.5   SPLITTING THE CHOSEN STRING

Next, we'll look at an example where there are many possible ways to split the string.

Theorem 11.5

The language $L = \{01^n 01^n : n \geq 0\}$ is not regular.

For this problem, the choice of string w is easy since we can just select one of the given form (here, $01^n 01^n$). Pause here to try to figure out what are the different ways to split the string into x, y, z such that $|y| > 0$ and $|xy| \leq n$ before reading on.

Hopefully, you found the edge case where $x = \lambda$. For this case, y must start with a 0 and have some number of 1's. Note that it is not possible for y to have a second 0 since that would mean that $|xy| > n$. The second case is that x has the first 0. We could further break this into the cases where x is just 0 and x is a 0 followed by some 1s, but this is more cases than we would need. If we let $x = 01^i$ where $i \geq 0$, then we cover both these sub-cases together. We can now write the proof.

Proof

Let us assume for a contradiction that L is regular. Then the pumping lemma applies to it for some $n > 0$. We pick the string $w = 01^n01^n$ and note that $w \in L$ and that $|w| = 2n + 2 > n$.

When we consider all possible ways to break up w into $w = xyz$ such that $|y| > 0$ and $|xy| \leq n$ there are two cases.

Case 1: ($x = \lambda$, $y = 01^i, 0 \leq i < n$, $z = 1^{n-i}01^n$) In this case, if we pump the string $k = 0$ times, then we get the string $xz = 1^{n-i}01^n$, which has only one 0 and thus cannot be in the language L.

Case 2: ($x = 01^i$, $y = 1^j$, $0 \leq i < n$, $1 \leq j < n$, $z = 1^{n-i-j}01^n$) In this case, if we pump the string $k = 2$ times, then the pumped string $xyyz = 01^i1^{2j}1^{n-i-j}01^n = 01^{n+j}01^n$ is not in the language as $n + j \neq n$ since $j > 0$.

Since we arrived at a contradiction for all possible ways to legally split the string w, this contradicts the pumping lemma. Thus, L is not regular.

Notice that it is fine to pick different values of k in the different sub-cases.

11.6 CHOOSING THE NUMBER OF TIMES TO PUMP

The next example we will look at has us be careful about the k that we select.

Theorem 11.6

The language $L = \{0^n1^m : n > m\}$ is not regular.

Proof

Let us assume for a contradiction that L is regular. Then the pumping lemma applies to it for some $n > 0$. We pick $w = 0^{n+1}1^n$ and note that $w \in L$ because the number of 0s is larger than the number of 1s and $|w| = 2n + 1 > n$.

We now consider all possible ways to break up w into $w = xyz$ such that $|y| > 0$ and $|xy| \leq n$. Since the first n symbols in w are all 0s, the strings x and y can only have 0s in them.

Let us say that $x = 0^i$ and $y = 0^j$, where $i \geq 0$ and $j \geq 1$. This means that the remaining symbols in w go into $z = 0^{n+1-i-j}1^n$.

We have to be careful in our choice of k. We pick $k = 0$ and get that $xy^0z = xz = 0^i0^{n+1-i-j}1^n = 0^{n+1-j}1^n$, which is not in L since $n + 1 - j \leq n$ as $j \geq 1$. This contradicts the pumping lemma. Thus, L is not regular.

Notice in this example that we had to be very particular about the k that we chose. Here, only $k = 0$ would work as any larger k would pump the string into a string in L.

11.7 A MORE COMPLEX EXAMPLE

This final example shows how we have to be careful to select a string in the language and be creative in how we pump the string.

> **Theorem 11.7**
>
> The language $L = \{0^n : n$ is a perfect square$\}$ is not regular.

In the pumping lemma proof for this language, when we select a string from the language, it is tempting to pick a string such as 0^n. However, note that we know nothing about n and in particular we don't know if it is a perfect square. To make the conditions of our selection of w work, we need a string of length a perfect square, so we ensure this by picking $w = 0^{n^2}$.

> **Proof**
>
> Let us assume for a contradiction that L is regular. Then the pumping lemma applies to it for some $n > 0$. We pick $w = 0^{n^2}$ and note that n^2 is a perfect square and that $n^2 \geq n$.
>
> Next, we consider all possible ways to break up w into $w = xyz$ such that $|y| > 0$ and $|xy| \leq n$. Since all the symbols in w are 0s, this part is easy. Let us say that $x = 0^i$ and $y = 0^j$, where $0 \leq i < n$ and $1 \leq j \leq n$. This means that the remaining symbols in w go into $z = 0^{n^2 - i - j}$.
>
> We now have to pick a k so that the pumped string is not in the language. To accomplish this, we pump the string few enough times so that we don't reach the next larger perfect square $((n+1)^2)$. Selecting $k = 2$, we get that $xy^k z = xyyz = 0^i 0^{2j} 0^{n^2 - i - j} = 0^{n^2 + j}$. Now, since $j > 0$ we know that $n^2 + j$ is strictly bigger than n^2. We now use the fact that $j \leq n$ (since we assumed that $|xy| \leq n$) to see that $n^2 + j \leq n^2 + n$ is strictly smaller than $n^2 + 2n + 1 = (n+1)^2$. Since $n^2 + j$ is between n^2 and $(n+1)^2$, it cannot be a perfect square.
>
> This contradicts the pumping lemma. Thus, the language L is not regular.

To make this proof work, we had to come up with a way of showing that the pumped string did not have a length that was a perfect square. To do this, we had to make use of all the conditions ($|y| > 0$ and $|xy| \leq n$) to show that the pumped string fell between two perfect squares.

11.8 CHAPTER SUMMARY AND KEY CONCEPTS

- The **pigeonhole principle** states that if more than n objects are assigned to n containers, then some container must have more than one of the objects.

- The **pumping lemma** demonstrates a property that must be true for all regular languages.

- We can show that a language is not regular by assuming for a contradiction that it is regular and proving that the pumping lemma does not apply to it.

EXERCISES

Prove that the following languages are not regular:

11.1 $\{0^n 1^{2n} : n \geq 0\}$

11.2 $\{0^n 1^m : n < m\}$

11.3 $\{0^n 1^m : n \leq m\}$

11.4 $\{0^n 1^m : n \geq m\}$

11.5 $\{0^n 1^m 0^n 1^m : n, m \geq 0\}$

11.6 $\{0^n 1^m 0^m 1^n : n, m \geq 0\}$

11.7 $\{0^n 1^m 0^{n+m} : n, m \geq 0\}$

11.8 $\{0^n 1^m 0^{nm} : n, m \geq 0\}$

11.9 $\{0^n 1^m 0^l : n, m, l \geq 0, l \leq n + m\}$

11.10 $\{0^n 1^m 0^l : n, m, l \geq 0, l \geq n + m\}$

11.11 The set of strings of matched parentheses over $\{(,)\}$ (e.g., $(())$ and $()((()))$ are in the language, but $())$ is not)

11.12 $\{ww : w \in \{0,1\}^*\}$

11.13 $\{ww^R : w \in \{0,1\}^*\}$ (Note: w^R is the reverse of the string w.)

11.14 $\{(01)^n (10)^n : n \geq 0\}$

11.15 $\{w \in \{0,1\}^* : w \text{ is a palindrome}\}$ (Note: A string is a palindrome if it is the same read forwards as backwards, e.g., 0, 101, 1001, 111010111.)

11.16 $\{0^n : n \text{ is a perfect cube}\}$

11.17 $\{0^n : n \text{ is a perfect fourth power}\}$

11.18 $\{0^n : n \text{ is a power of 2}\}$

11.19 (Harder) $\{0^n : n \text{ is a prime number}\}$

11.20 (Harder) $\{w \in \{0,1\}^* : w \text{ is not a palindrome}\}$

CHAPTER 12

Context-Free Grammars

12.1 WHY YOU SHOULD CARE

Now that we have established that DFAs are not powerful enough to capture all computation, we will look at a different way to model languages called Context-Free Grammars (CFGs). These grammars lay the foundation for all natural languages, such as English, Hindi, Spanish, and Vietnamese and programming languages, such as C++, Java, and Python.

12.2 AN EXAMPLE CONTEXT-FREE GRAMMAR

We'll start with an example of a context-free grammar (CFG) for the non-regular language $L = \{0^n 1^n : n \geq 0\}$.

> **Example 12.1**
>
> A grammar is primarily defined by its productions:
>
> $$S \to 0S1$$
> $$S \to \lambda$$

In the above example S is called a variable or non-terminal. In particular, S is what is called the initial variable. Each production of the grammar has a variable on the left and an expression on the right that is any combination of variables, symbols from the alphabet Σ (called terminals), and λ. Starting from the initial variable, we can produce strings by applying productions until we have only terminals left. For example, consider how we would produce strings using the above grammar:

> **Example 12.2**
>
> We can generate 0011 as:
>
> $$S \Rightarrow 0S1 \Rightarrow 00S11 \Rightarrow 0011$$
>
> by replacing S with $0S1$, $0S1$, and λ in the three steps.

 DOI: 10.1201/9781003383284-12

Example 12.3

The string 000111 can be generated as:

$$S \Rightarrow 0S1 \Rightarrow 00S11 \Rightarrow 000S111 \Rightarrow 000111.$$

Example 12.4

Even the string λ (when $n = 0$) can be generated as:

$$S \Rightarrow \lambda.$$

It should be clear that we can similarly apply the $S \to 0S1$ production n times (for any $n \geq 0$), followed by the $S \to \lambda$ production to get any string in $L = \{0^n 1^n : n \geq 0\}$. Even the empty string can be derived (as it should, since $0^0 1^0 = \lambda$) by right away applying the $S \to \lambda$ rule. Moreover, because of the limited productions, we cannot produce any strings not in L. Thus, the example grammar does indeed generate precisely the strings in L.

12.3   PALINDROMES

We will next see an example of a CFG for the language $P = \{w \in \{0,1\}^* : w$ is a palindrome$\}$. Recall that a palindrome is a string that reads the same forwards and backwards, such as the word *racecar*. We will show how to make a grammar for this language over the binary alphabet $\Sigma = \{0,1\}$.

Example 12.5

The productions of a CFG for P are:

$$S \to 0S0$$

$$S \to 1S1$$

$$S \to 0$$

$$S \to 1$$

$$S \to \lambda$$

To abbreviate this notation, the productions of a grammar can alternatively be given with the | symbol used to separate different right hand sides as below:

$$S \to 0S0 \mid 1S1 \mid 0 \mid 1 \mid \lambda$$

You should verify that any palindrome over $\{0,1\}$ can be generated by this grammar and that this grammar only generates palindromes. A few example derivations are given next:

Example 12.6

For the string 101101:

$$S \Rightarrow 1S1 \Rightarrow 10S01 \Rightarrow 101S101 \Rightarrow 101101.$$

For the string 00100:

$$S \Rightarrow 0S0 \Rightarrow 00S00 \Rightarrow 00100.$$

For the string 0:

$$S \Rightarrow 0.$$

12.4 CONTEXT-FREE GRAMMARS FOR REGULAR LANGUAGES

You will show in an exercise that any regular language has a context-free grammar. Some examples of this are given next:

Example 12.7

The language of the regular expression $1(0+11)^*$ has the following grammar:

$$S \to 1A$$

$$A \to 0A \mid 11A \mid \lambda$$

Note here that we added a second variable, A, that generates the strings of the regular expression $(0+11)^*$. The strings in the language are the ones that are generated starting from S, the initial variable.

Here are how some strings are generated from this grammar:

Example 12.8

For the string 1011:

$$S \Rightarrow 1A \Rightarrow 10A \Rightarrow 1011A \Rightarrow 1011.$$

For the string 11111:

$$S \Rightarrow 1A \Rightarrow 111A \Rightarrow 11111A \Rightarrow 11111.$$

For the string 1110:

$$S \Rightarrow 1A \Rightarrow 111A \Rightarrow 1110A \Rightarrow 1110.$$

For the string 1:

$$S \Rightarrow 1A \Rightarrow 1.$$

Example 12.9

The language of the regular expression $1^*0^* + 0^*1^*(1 + 00)$ has this grammar:

$$S \to A \mid B$$

$$A \to CD$$

$$C \to 1C \mid \lambda$$

$$D \to 0D \mid \lambda$$

$$B \to DCE$$

$$E \to 1 \mid 00$$

Here, A, B, C, D, E generate the strings in the languages of 1^*0^*, $0^*1^*(1+00)$, 1^*, 0^*, $(1 + 00)$ respectively.

Example 12.10

For the string 110:

$$S \Rightarrow A \Rightarrow CD \Rightarrow 1CD \Rightarrow 11CD \Rightarrow 11D \Rightarrow 110D \Rightarrow 110.$$

For the string 0100:

$$S \Rightarrow B \Rightarrow DCE \Rightarrow 0DCE \Rightarrow 0CE \Rightarrow 01CE \Rightarrow 01E \Rightarrow 0100.$$

For the string 1:
$$S \Rightarrow A \Rightarrow CD \Rightarrow 1CD \Rightarrow 1D \Rightarrow 1,$$

or

$$S \Rightarrow B \Rightarrow DCE \Rightarrow CE \Rightarrow E \Rightarrow 1.$$

Note in this last example that there were multiple ways to derive the string.

12.5   FORMAL DEFINITION OF CFGS

Now that we have seen a few different grammars, we can formally define them.

Definition 12.1

A grammar G is defined by the tuple

$$G = (V, \Sigma, P, S),$$

where V is a set of variables or non-terminals, Σ is the alphabet, P is a set of productions, and S is the initial or start variable.

Example 12.11

The grammar in Example 12.9 can be represented by (V, Σ, P, S), where

$$V = \{S, A, B, C, D, E\}$$

$$\Sigma = \{0, 1\}$$

and

$$P = \{S \rightarrow A, S \rightarrow B, A \rightarrow CD, C \rightarrow 1C, C \rightarrow \lambda,$$
$$D \rightarrow 0D, D \rightarrow \lambda, B \rightarrow DCE, E \rightarrow 1, E \rightarrow 00\}.$$

The language of a CFG $G = (V, \Sigma, P, S)$ is defined as the set of strings that can be produced from S using productions in P until there are only terminals left. Any language that has a context-free grammar is called *context-free*.

Definition 12.2

A **context-free language** is any language that has a context-free grammar.

12.6   CLOSURE UNDER UNION

We can construct a CFG for the union of two context-free languages such as

$$U = \{0^n : n \geq 1\} \cup \{w01 : w \in \{0, 1\}^*\}$$

from the CFGs for the two individual languages.

Example 12.12

The productions of a CFG for U are:

$$S \rightarrow A \mid B$$

$$A \rightarrow 0A \mid 0$$

$$B \rightarrow C01$$

$$C \rightarrow 0C \mid 1C \mid \lambda$$

Here, the variable A can produce any string in $\{0^n : n \geq 1\}$, the variable C can produce any string over $\{0, 1\}$ and thus the variable B can produce any string that ends in 01, or $\{w01 : w \in \{0, 1\}^*\}$. Since S can produce strings from either A or B, S produces the language U.

We can generalize the above example to give a procedure to find a CFG for the union of any two context-free languages.

Theorem 12.1

If L_1 and L_2 are context-free languages, then $L_1 \cup L_2$ is a context-free language.

Proof

Let L_1 and L_2 be context-free languages. Then, by the definition of context-free languages, they must have grammars, say $G_1 = (V_1, \Sigma_1, P_1, S_1)$ and $G_2 = (V_1, \Sigma_1, P_1, S_1)$. We will give a construction for a new grammar that has all the strings, and only the strings, in $L_1 \cup L_2$.

The new grammar that we construct will use the productions from the grammars G_1 and G_2. However, before doing so, we should make sure that there are no overlapping variables in G_1 and G_2 as this could make the construction fail. If there are any variables in $V_1 \cap V_2$, then we can simply rename them to make them unique. Since the name of a variable doesn't affect the grammar in any way, this has no effect on the language of the grammars.

Let S be a new variable that is in neither V_1 nor V_2. We create the following new grammar $G = (V, \Sigma, P, S)$:

- $V = \{S\} \cup V_1 \cup V_2$

- $\Sigma = \Sigma_1 \cup \Sigma_2$

- $P = \{S \to S_1, S \to S_2\} \cup P_1 \cup P_2$

Since this new grammar has productions $S \to S_1$ and $S \to S_2$, it can generate any string from G_1 or from G_2. Moreover, these are the only strings that it can generate. Hence, G generates the language $L_1 \cup L_2$. Since $L_1 \cup L_2$ has a context-free grammar, it must be a context-free language as well.

You can show closure under concatenation and Kleene star in a similar manner. These are left as exercises.

12.7 APPLICATIONS OF CFGS

Grammars can be developed for a variety of languages, both computing languages and natural languages that are used by people.

Example 12.13

A programming language can define positive integers with the following productions:
$$N \to N0 \mid N1 \mid N2 \mid \ldots \mid N9 \mid 1 \mid 2 \mid \ldots \mid 9$$

We can then get all integers with the rules:

$$I \to 0 \mid N \mid -N$$

Example 12.14

Arithmetic expressions over integers can then be given by the grammar:

$$E \to I \mid E + E \mid E - E \mid E * E \mid E/E \mid E\%E \mid (E)$$

Example 12.15

We can build on the previous examples to write a grammar for an entire programming language in the following manner:

$$PROGRAM \to PROGRAM\ STATEMENT$$

$$STATEMENT \to IFSTATEMENT \mid FORSTATEMENT \mid \ldots$$

Indeed, all programming languages have grammars defined for them.

In a similar way, we can write grammars for languages used by people. Computer scientists call these natural languages.

Example 12.16

A grammar for sentences in the English language might look something like this:

$$SENTENCE \to NOUNPHRASE\ VERBPHRASE$$

$$NOUNPHRASE \to ARTICLE\ ADJECTIVE\ NOUN$$

$$ARTICLE \to the \mid a \mid an \mid \ldots$$

$$ADJECTIVE \to ADJECTIVE\ ADJECTIVE \mid brown \mid happy \mid black \mid \ldots$$

$$NOUN \to dog \mid cat \mid \ldots$$

$$VERBPHRASE \to VERB\ NOUNPHRASE$$

$$VERB \to saw \mid ate \mid \ldots$$

Such a grammar would allow us to produce declarative sentences such as "the happy brown dog saw the black cat." As an exercise, figure out how you would generate this sentence using the above grammar. This action is known as parsing.

While CFGs have many useful applications in computer science, they cannot capture all computable languages. For example, it is possible to use a pumping lemma for context-free languages to show that there can be no context-free grammar for $\{0^n 1^n 2^n : n \geq 1\}$. Thus, we will need a more powerful model, such as the Turing machine that you will learn about in the next chapter.

12.8   CHAPTER SUMMARY AND KEY CONCEPTS

- A **context-free grammar** can be used to generate strings. The set of strings generated by the grammar are called its language.

- Context-free grammars can be found for **all regular languages**.

- The **formal definition** of a context-free grammar is given by $G = (V, \Sigma, P, S)$, where V is a set of **variables**, Σ is the **alphabet**, P is a set of **productions**, and S is the **start variable**.

- Any language that has a context-free grammar is called a **context-free language**.

- Context-free languages are **closed under union**.

- Context-free grammars have computer science applications in the subareas of **programming languages** and **natural language processing**.

EXERCISES

Give grammars for the following languages:

12.1 $\{0^n 1^{2n} : n \geq 0\}$

12.2 $\{(01)^n (10)^n : n \geq 1\}$

12.3 $\{0^n 1^m : n \leq m\}$

12.4 $\{0^n 1^m : n \geq m\}$

12.5 $\{01^n 01^n : n \geq 0\}$

12.6 $\{0^n 1^n 0^m 1^m : n, m \geq 0\}$

12.7 $\{0^n 1^m 0^m 1^n : n, m \geq 0\}$

12.8 $\{0^n 1^m 0^{n+m} : n \geq 0\}$

12.9 $\{0^n 1^m 0^{2n+m} : n \geq 0\}$

12.10 $\{0^n 1^m 0^l : l < n + m\}$

12.11 $\{0^n 1^m 0^l : l > n + m\}$

12.12 $\{0^n 1^m 0^l : l \neq n + m\}$

12.13 $\{1^m : \text{ is a multiple of 3 or 5}\}$

12.14 $(0 + 1)^*010(0 + 1)^*$

12.15 The set of strings of matched parentheses over $\{(,)\}$ (e.g., $(())$ and $()((()))$ are in the language, but $())$ is not)

12.16 The set of strings of matched parentheses over $\{(,),[,],\{,\}\}$ (e.g., $([])$ and $()\{[()]\}$ are in the language, but $(])$ is not)

12.17 Show that it was important to make sure that the variables of the two grammars were unique in the proof of Theorem 12.1 by giving an example in which they are not unique and the construction does not give the union of the two languages.

12.18 Use the construction in the proof of Theorem 12.1 to give a grammar for

$$\{0^n 1^n : n \geq 0\} \cup \{w : w \text{ is a palindrome over } \{0,1\}\}.$$

12.19 Show that the context-free languages are closed under concatenation. That is, show that if L_1 and L_2 are context-free languages, then

$$L = \{s_1 s_2 : s_1 \in L_1 \wedge s_2 \in L_2\}$$

is a context-free language.

12.20 Show that the context-free languages are closed under Kleene star. That is, show that if L is a context-free languages, then

$$L' = \{s_1 s_2 s_3 \ldots s_n : n \geq 0 \wedge s_1, \ldots, s_n \in L\}$$

is a context-free language.

12.21 Prove using strong induction that every regular language has a grammar. (Hint: You can use the closure under union proof from this chapter for part of this proof.)

CHAPTER 13

Turing Machines

13.1   WHY YOU SHOULD CARE

The Turing machine is the standard model for all computation. It is capable of accepting all regular languages, context-free languages, and indeed any language computation that is believed to be possible by a computer. All programming languages are called Turing-complete, which is an informal way of saying that they can do all that a Turing machine can do. Thus understanding how Turing machines work is important for any computer scientist.

13.2   AN EXAMPLE TURING MACHINE

What makes the DFA an unsatisfactory model of computation is that it necessarily has finite memory and thus cannot count arbitrarily high, leading it to be unable to accept simple languages like

$$L = \{0^n 1^n : n \geq 0\}.$$

To get around this deficiency, the Turing machine adds an infinitely long tape (stretching forever to the left and right) that has data on it in the form of cells, each of which holds a single character and can be modified by a head that moves along the tape one cell at a time. This unbounded memory allows us to model all known computation. The input string now starts on the tape with the head pointed to the leftmost character. All other cells are filled with a special blank symbol (usually B).

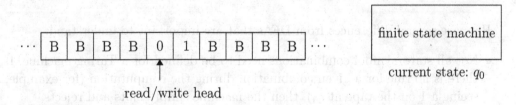

Figure 13.1 A Turing machine with the input 01 on its doubly infinite tape

DOI: 10.1201/9781003383284-13

Example 13.1

Turing machine for $\{0^n 1^n : n \geq 0\}$:

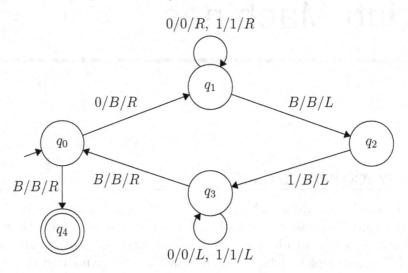

As with a DFA, we start at a start state (q_0 in this example) and transition between states. The difference now is that we transition based on the state and the character in the cell at the head position. Moreover, we have to specify what gets written to the cell and whether the head should move to the left (denoted by L) or the right (denoted by R). For example, notice that the state q_0 has an arrow out of it with $0/B/R$ to q_1. This means that if the Turing machine is at state q_0 and it sees a 0 on the tape at its head position, it will write a B at that position, move the head to the right, and switch to state q_1. Equipped with this understanding, we can now follow the algorithm being executed by the Turing machine above. The machine will erase (replace with a B) the first 0 and move to state q_1 which will traverse the rest of the string. When the end of the string is reached (indicated by a B to the right), we switch to state q_2. From q_2, the rightmost 1 is erased and we move to q_3 which moves the head back to the beginning of the string. Continuing in this manner, we can erase 0s and 1s from the two ends of the string. If the string is really in the language, then we will eventually erase the entire string at which point we follow the transition from q_0 to q_4 on a blank and accept.

Here are several differences from DFAs that are important to understand:

• Not all state/symbol combinations need to be defined for a Turing machine. If there is no rule for a given combination during the computation (for example, seeing a 1 on the tape at q_0), then the machine simply halts and rejects.

• The transitions that are defined are still deterministic—we don't allow a state/symbol combination to appear multiple times in this model of the Turing machine. There are non-deterministic models of Turing machines, but we will not study them in this book.

- Unlike DFAs for which we go through the string only once, for a Turing machine we can traverse the string multiple times and go back to characters we have processed previously. In fact, most interesting problems will require us to make several passes over the input or use the blank tape cells in their computation.

- A Turing machine accepts a string simply by reaching an accepting state. As soon as an accepting state is reached, the computation halts and accepts.

In particular, the last item means that we do not have to "consume" the entire string to accept/reject it. For example, a Turing machine for Σ^* can have a single accepting state (and no others) or a Turing machine for $\emptyset$ can be one with a single starting state that is not accepting (and no others).

We will next see some notation for tracing the steps of a Turing machine. The trace will contain all the characters on the tape (other than the blank symbols that go off infinitely to the left and right) with the current state written immediately to the left of the position of the head.

Example 13.2

For example, the trace for the earlier Turing machine on the input 0011 is given below:

$q_0 0011 \Rightarrow q_1 011 \Rightarrow 0q_1 11 \Rightarrow 01q_1 1 \Rightarrow 011q_1 B \Rightarrow 01q_2 1 \Rightarrow 0q_3 1 \Rightarrow q_3 01 \Rightarrow$
$q_3 B01 \Rightarrow q_0 01 \Rightarrow q_1 1 \Rightarrow 1q_1 B \Rightarrow q_2 1 \Rightarrow q_3 B \Rightarrow q_0 B \Rightarrow q_4 B$ (ACCEPT)

Example 13.3

The trace for the earlier Turing machine on the input 000111 starts out as below:

$q_0 000111 \Rightarrow q_1 00111 \Rightarrow 0q_1 0111 \Rightarrow 00q_1 111 \Rightarrow 001q_1 11 \Rightarrow 0011q_1 1 \Rightarrow 00111q_1 B$
$\Rightarrow 0011q_2 1 \Rightarrow 001q_3 1 \Rightarrow 00q_3 11 \Rightarrow 0q_3 011 \Rightarrow q_3 0011 \Rightarrow q_3 B0011 \Rightarrow q_0 0011 \Rightarrow \ldots$

At this point, the state of the Turing machine is identical to that in the starting point in the previous example and the computation goes on to accept.

Example 13.4

On the other hand, the string 011 will not be accepted:

$q_0 011 \Rightarrow q_1 11 \Rightarrow 1q_1 1 \Rightarrow 11q_1 B \Rightarrow 1q_2 1 \Rightarrow q_3 1 \Rightarrow q_3 B1 \Rightarrow q_0 1$ (REJECT)

The language of a Turing machine is the set of strings accepted by it. Notice that this first example Turing machine will halt and accept or halt and reject on all strings, but it is possible that a Turing machine can reject by running forever. This detail will be important later.

13.3 FORMAL DEFINITION OF A TURING MACHINE

A Turing machine can be defined using a 7-tuple as follows.

Definition 13.1

Any Turing machine can be defined by a tuple of the form $(Q, \Sigma, \Gamma, \delta, q_0, B, F)$, where

- Q is the set of states,
- Σ is the input alphabet,
- Γ (upper case gamma) is the tape alphabet (note: $\Sigma \subseteq \Gamma$ and $B \in \Gamma$),
- $\delta : Q \times \Gamma \to Q \times \Gamma \times \{L, R\}$ is the transition function—this takes as inputs the current state and the tape alphabet the head is pointing to and outputs the next state, what to write at the head position, and whether to move left (L) or right (R),
- q_0 is the start state,
- B is the special blank symbol, and
- F is the set of accepting states.

Example 13.5

For the Turing machine in Example 13.1, the tuple $(Q, \Sigma, \Gamma, \delta, q_0, B, F)$ is given by

- $Q = \{q_0, q_1, q_2, q_3, q_4\}$,
- $\Sigma = \{0, 1\}$,
- $\Gamma = \{0, 1, B\}$,
- δ is the transition function from the figure ($\delta(q_0, 0) = (q_1, B, R)$, $\delta(q_1, 0) = (q_1, 0, R), \ldots$),
- q_0 is the starting state,
- B is the blank symbol, and
- $F = \{q_4\}$.

13.4 RECOGNIZING ADDITION

Our second example will demonstrate how Turing machines can recognize addition with unary numbers. Unary, or tally numbering, is a very simple numerical encoding that has a single symbol (usually 1 or 0) and each positive integer is represented by that number of that symbol.

Example 13.6

In unary, the number 4 is represented by 1111 and 7 is represented by 1111111.

The following language allows us to recognize or check addition in unary by using 0s for the first number, 1s for the second, and 0s for the sum:

$$L_+ = \{0^n 1^m 0^{n+m} : n, m \geq 0\}.$$

To design a Turing machine for this language, we can use an algorithm similar to the previous example. We can match 0s at the start of the string with 0s at the end by erasing them. Once this is done we erase matched pairs of 1s from the beginning with 0s on the end until the entire string is erased. If the end result of the computation is a blank tape, then the computation should accept. If the string is not of the given form, the computation should reject.

Example 13.7

A Turing machine for

$$L_+ = \{0^n 1^m 0^{n+m} : n, m \geq 0\}$$

is given in Figure 13.2. This Turing machine executes the algorithm outlined earlier. The states q_0, q_1, q_2, q_3 erase the leftmost 0 and the rightmost 0 repeatedly until there are no more 0s at the start of the string. The states q_4, q_5, q_6, q_7 then in a similar fashion erase the leftmost 1 and the rightmost 0 repeatedly. If the input string is of the correct form, then the entire string will be erased and the computation will accept. In all other cases it will reject.

Note that this Turing machine correctly handles special cases such as strings of the form $0^n 1^m 0^{n+m}$, where $n = 0$, $m = 0$, and both $n = 0$ and $m = 0$. Trace through examples of these cases to see why this is true. As with coding, you want to pay attention to special cases like these and make sure your algorithm is working correctly for them.

Example 13.8

The trace for the above Turing machine on the input 0100 (demonstrating that $1 + 1 = 2$) looks like this:

$q_0 0100 \Rightarrow q_1 100 \Rightarrow 1 q_1 00 \Rightarrow 10 q_1 0 \Rightarrow 100 q_1 B \Rightarrow 10 q_2 0 \Rightarrow 1 q_3 0 \Rightarrow q_3 10$

$\Rightarrow q_3 B10 \Rightarrow q_0 10 \Rightarrow q_4 0 \Rightarrow 0 q_4 B \Rightarrow q_5 0 \Rightarrow q_6 B \Rightarrow q_7 B \Rightarrow q_8 B$ (ACCEPT)

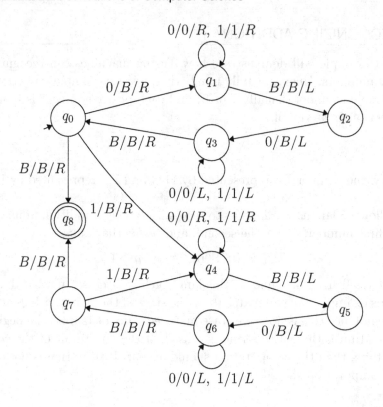

Figure 13.2 A Turing machine for $\{0^n 1^m 0^{n+m} : n, m \geq 0\}$.

Example 13.9

The rejecting trace on the input 0011000 (demonstrating that $2 + 2 \neq 3$) looks like this:

$q_0 0011000 \Rightarrow q_1 011000 \Rightarrow 0q_1 11000 \Rightarrow 01q_1 1000 \Rightarrow 011q_1 000 \Rightarrow 0110q_1 00$

$\Rightarrow 01100q_1 0 \Rightarrow 011000q_1 B \Rightarrow 01100q_2 0 \Rightarrow 0110q_3 0 \Rightarrow 011q_3 00 \Rightarrow 01q_3 100$

$\Rightarrow 0q_3 1100 \Rightarrow q_3 01100 \Rightarrow q_3 B01100 \Rightarrow q_0 01100 \Rightarrow q_1 1100 \Rightarrow 1q_1 100 \Rightarrow 11q_1 00$

$\Rightarrow 110q_1 0 \Rightarrow 1100q_1 B \Rightarrow 110q_2 0 \Rightarrow 11q_3 0 \Rightarrow 1q_3 10 \Rightarrow q_3 110 \Rightarrow q_3 B110$

$\Rightarrow q_0 110 \Rightarrow q_4 10 \Rightarrow 1q_4 0 \Rightarrow 10q_4 B \Rightarrow 1q_5 0 \Rightarrow q_6 1 \Rightarrow q_6 B1 \Rightarrow q_7 1 \Rightarrow q_4 B$

$\Rightarrow q_5 B$ (REJECT)

You should also verify that the Turing machine works correctly when $n = 0$ (for example, by accepting strings such as 10 and 1100 and rejecting strings such as 110 and 100), when $m = 0$ (for example, by accepting strings such as 00 and 0000 and rejecting strings such as 0 and 000), and when both n and m are zero (i.e., when the input is λ).

13.5 CONDITIONAL BRANCHING WITH A TURING MACHINE

We will next see an example of a Turing machine for a more complex language, the set of strings with equal numbers of as and bs. This example will illustrate the idea of branching with a Turing machine. The algorithm this time will mark off the leftmost a or b with a new symbol and then move to the right through the string to find a matching complementary symbol (b or a, respectively). Once this has been marked off, the Turing machine returns to the start of the string and repeats this procedure again. The computation terminates successfully when all the as and bs have been marked off. If this does not happen, then the computation rejects.

Here is a Turing machine for the language

$$L_= = \{w \in \{a, b\}^* : w \text{ has an equal number of a's and b's}\}.$$

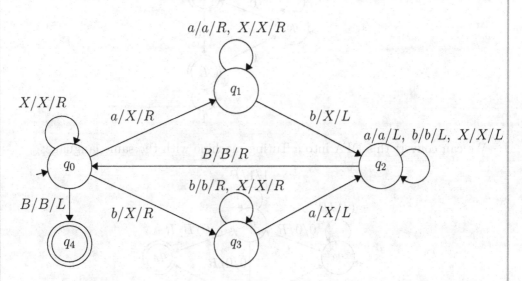

This example introduces a new symbol, X, that is not in the input alphabet but is in the tape alphabet. The Turing machine works by branching depending on whether the first character is an a or a b. If it is an a, it replaces it with an X and takes the upper branch (q_1) to move to the right until it finds a b that can be replaced with an X. If it is a b, it also replaces this with an X and takes the lower branch (q_3) to move to the right until it finds an a that can be replaced with an X. In both cases, the state q_2 will then move the head to the left until it finds the start of the string and repeats the procedure again (from q_0) for the next a or b that has not been replaced with an X. If all the characters in the string have been replaced by Xs, the Turing machine goes to the accepting state. Again, note that the special case of the empty string is handled correctly as we accept when we see a B on q_0.

13.6 TURING MACHINES CAN ACCEPT ALL REGULAR LANGUAGES

We will show that every regular language has a Turing machine that accepts it by simulating the actions of the DFA with its finite state. We'll start with an example and then generalize this idea to any regular language.

Example 13.11

The regular language

$$\{s \in \{0,1\}^* : \text{ the number of 0s in } s \text{ is not divisible by 3}\}$$

is accepted by the following DFA:

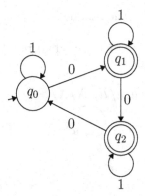

We can convert this DFA into a Turing machine with the same language:

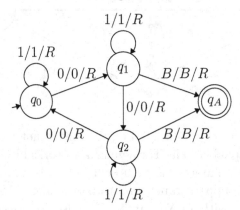

Notice that (1) we changed each DFA transition into a Turing machine transition with the same states and symbol, leaving the tape unchanged and always moving right, (2) we made the accepting states of the DFA into non-accepting ones (as this might lead to a premature acceptance from the Turing machine), and (3) we added a new accepting state q_A that we transition to from each accepting state (q_1 and q_2 in this example) on a B since we should accept precisely the strings that end at an accepting state and a B indicates the end of the string.

Theorem 13.1

Every regular language has a Turing machine that accepts it.

Proof

Let L be any regular language. By the definition of regular languages, L must have a DFA for it $D = (Q, \Sigma, \delta, q_0, F)$. We will generate a Turing machine that accepts the same language as D.

Define the Turing machine $T = (Q \cup \{q_A\}, \Sigma, \Sigma \cup \{B\}, \delta', q_0, B, \{q_A\})$ as follows:

- For each transition $\delta(q_s, c) = q_d$ ($q_s, q_d \in Q, c \in \Sigma$) for the DFA, add the transition $\delta'(q_s, c) = (q_d, c, R)$ for the Turing machine.

- For each accepting state of the DFA $q \in F$, add the transition $\delta'(q, B) = (q_A, B, R)$ for the Turing machine.

The behavior of the constructed Turing machine mimics that of the DFA precisely since all the head moves are to the right, iterating through the input string starting from the start state. Moreover, once the end of the string is reached (signaled with a B symbol), the Turing machine has a transition to its accepting state precisely when the string would have been accepted by the DFA (i.e., from one of the accepting states of the DFA). Thus, the Turing machine accepts the language of the DFA. This means that the regular language L also has a Turing machine that accepts it.

13.7 TURING MACHINES AS COMPUTERS OF FUNCTIONS

Besides computing languages, Turing machines can also compute functions. Specifically, such a the Turing machine starts with its input on the tape and writes the output of the function before it halts.

Example 13.12

Turing machine to increment a binary counter:

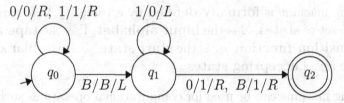

This Turing machine works by moving to the rightmost character of the string and then replacing 1s with 0s until it encounters a 0 that it changes to a 1. A special case is if it encounters no 0s, then it carries a 1 to the next blank cell to the left.

Example 13.13

For example, the previous Turing machine will increment from 1000 to 1001, from 1011 to 1100, and from 1111 to 10000.

The next simple example shows how to add two unary numbers (as defined in Section 13.4), represented by 0^n and 1^m. To compute the sum for 0^n1^m we need to create the output 1^{n+m}. This is very simple to do by converting all the initial 0s to 1s as in the following example:

Example 13.14

Turing machine to add two unary numbers:

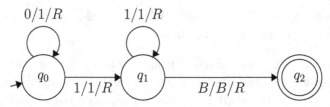

Example 13.15

For example, the previous Turing machine will convert the input 00111 into 11111, showing that it can add $2 + 3$ to get 5.

13.8 CHAPTER SUMMARY AND KEY CONCEPTS

- The **Turing machine** is the standard model for all possible computation. Any computational system that can do all that a Turing machine does is called **Turing-complete**.

- A Turing machine has a **doubly-infinite tape** with a read/write **tape head** that can move along the tape. It also has **finite control** in the form of states with transitions.

- A Turing machine is **formally defined** by a tuple $(Q, \Sigma, \Gamma, \delta, q_0, B, F)$, where Q is the set of **states**, Σ is the **input alphabet**, Γ is the **tape alphabet**, δ is the **transition function**, q_0 is the **start state**, B is the **blank symbol**, and F is the set of **accepting states**.

- A Turing machine can be used for computational operations such as **recognizing arithmetic operations** and performing **conditional branching**.

- Every regular language has a Turing machine that accepts it.

- Turing machines can also be used to **compute functions** with string inputs and outputs.

EXERCISES

Create Turing machines for the following languages:

13.1 $\{w \in \{0,1\}^* : w \text{ has } 01 \text{ as a contiguous substring}\}$

13.2 $\{w \in \{0,1\}^* : w \text{ ends in } 00\}$

13.3 $\{w \in \{0,\dots,9\}^* : w \text{ is even in decimal}\}$ (λ is even)

13.4 $\{w \in \{0,1\}^* : w \text{ has an odd number of 0s and an even number of 1s}\}$

13.5 $\{0^n 1^m : n \leq m\}$

13.6 $\{0^n 1^m : n \geq m\}$

13.7 $\{0^n 1^{2n} : n \geq 0\}$

13.8 $\{0^n 1^n 2^n : n \geq 0\}$

13.9 $\{0^n 1^n 0^m 1^m : n, m \geq 0\}$

13.10 $\{0^n 1^m 0^m 1^n : n, m \geq 0\}$

13.11 $\{(01)^n (10)^n : n \geq 1\}$

13.12 $\{0^n 1^m 0^{n-m} : n > m > 0\}$

13.13 $\{0^n 1^m 0^{2n+m} : n \geq 0\}$

13.14 $\{0^n 1^m 0^{nm} : n, m \geq 1\}$

13.15 even length palindromes over $\{0,1\}$

13.16 odd length palindromes over $\{0,1\}$

13.17 all palindromes over $\{0,1\}$

13.18 all non-palindromes over $\{0,1\}$

13.19 $\{w \in \{0,1,2\}^* : w \text{ has an equal number of 0s, 1s, and 2s}\}$

13.20 (Harder) $\{ww : w \in \{0,1\}^*\}$

13.21 (Harder) The set of strings of matched parentheses over $\{(,)\}$ (e.g., $(())$ and $()((()))$ are in the language, but $())$ is not)

13.22 (Harder) The set of strings of matched parentheses over $\{(,),[,],\{,\}\}$ (e.g., $([])$ and $()\{[()]\}$ are in the language, but $(]$ is not)

Create Turing machines for the following functions:

13.23 Given a number in binary as input, multiply the number by 2.

13.24 Given a number in binary as input, increment the number by 2.

13.25 Given a number in binary as input, decrement the number by 1 (or output 0 if the input is zero).

13.26 (Harder) Given a pair of numbers in binary separated by the special # tape character, output the sum of the numbers.

13.27 (Harder) Given a pair of numbers in binary separated by the special # tape character, output the difference of the numbers.

13.28 (Harder) Given a pair of numbers in binary separated by the special # tape character, output the result of modding the first by the second.

13.29 (Harder) Given a pair of numbers in binary separated by the special # tape character, output the product of the numbers.

13.30 (Harder) Given a pair of numbers in binary separated by the special # tape character, output the quotient of the numbers (rounding down).

CHAPTER 14

Computability

14.1 WHY YOU SHOULD CARE

In this chapter we will study what Turing machines are capable of from a higher level and see why we believe this model of computation to be the one that captures everything we believe a computer capable of doing. We will then show that there is a problem that is uncomputable. Since we established Turing machines as our model of computation, we will see that no mechanical computer that currently exists or ever will exist can solve this problem. We will then show that a number of other problems are also uncomputable by bootstrapping from this one problem using a technique called a reduction, another important concept in computer science.

14.2 VARIATIONS OF TURING MACHINES

While the standard model of the Turing machine from the last chapter is as powerful as we will need, we will see in this section that it is equivalent in power to seemingly more advanced models of computation, such as Turing machines that have multiple tapes.

We will first see how to construct a Turing machine with multiple tracks using a standard Turing machine. A two-track Turing machine has each cell on the tape split into a top and bottom part so that each cell can hold two symbols. (See Figure 14.1.) This model of the Turing machine can then make transitions based on the state and the symbols on both of the tracks.

We can simulate any such two-track Turing machine by treating the tape alphabet as pairs of symbols from the original tape alphabet. For example, if a cell had a 1 and a B in it, we would represent this with a single symbol $\frac{1}{B}$ and make a standard Turing machine that did whatever the two-track Turing machine did by designing the appropriate rules. Similarly, we can see that a multi-track Turing machine, for some number of tracks k, would be no more powerful than a standard Turing machine because we can replace the tape alphabet with k-tuples that represent any combination of k symbols from the original tape alphabet.

We can also design multi-tape Turing machines in which there are multiple independent tapes, for which each tape has a separate head. (See Figure 14.2.) Now, each head can be at a different cell of the tape. A k-tape Turing machine can be simulated

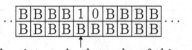

head points to both tracks of this cell

Figure 14.1 Multi-track TM

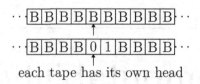

each tape has its own head

Figure 14.2 Multi-tape TM

by a $2k$-track Turing machine by using odd numbered tracks to represent the tapes and the even numbered tracks to maintain the head position of the track before it. We omit the details of how to design such a Turing machine as they would be excessively complex. (You can take a Theory of Computation course to learn more.) Since we know that we can simulate a $2k$-track Turing machine with a standard one, we can now simulate a Turing machine with any (finite) number of independent tapes with a standard Turing machine from the last chapter.

The point of demonstrating these variations is to show you that there are more complex models possible that are no more powerful than the Turing machine. These variations are also getting closer to models of computation you are used to in which you can keep track of multiple values at the same time.

14.3 THE CHURCH-TURING THESIS

Now that we have seen that Turing machines are capable of maintaining multiple tapes (variables in memory) and can perform many of the operations we deem necessary for computation (arithmetic, loops, conditionals) in examples and exercises in the last chapter, we are ready to make an assertion about the power of Turing machines. The following is one version of what is called the Church-Turing thesis, named after two pioneers in the theory of computation, Alonzo Church and Alan Turing.

Definition 14.1

The **Church-Turing thesis** is that all mechanical computation can be performed by a Turing machine.

The above statement is called a thesis because it is a hypothesis. Since the term "mechanical computation" is not well defined, the above thesis cannot be mathematically proved. It hypothesizes that any computation that we think is feasible can be done by a Turing machine. All modern computers are no more or less powerful than a Turing machine in that they can solve the same problems. Given enough time and memory, a Turing machine can perform any computation that we can do on a computer. Conversely, modern computers can perform any computation that a Turing machine is capable of given enough time and computer memory.

Now, this is not to say that we will never be able to build a computer that is more powerful than a Turing machine. It just means that we don't currently believe some more powerful computer is possible and all empirical evidence seems to support that belief.

14.4   UNIVERSAL TURING MACHINES

An important property of the Turing machine is that it is powerful enough to simulate itself. That is, given the encoding of a Turing machine and an input string, we can design a Turing machine (called a **Universal Turing Machine**) that can simulate the given Turing machine on the given input.

You might be wondering how you can encode a Turing machine as a string. Since the description of a Turing machine is always finite, we can encode the description into a single string. For example, we can encode the name of the states, the input and tape alphabet, the starting and accepting states, and the blank symbol into a string. We can then encode the δ transition rules as sequences—for example, $\delta(q_0, 0) = (q_1, B, R)$ could be encoded by a string such as $q0\#0\#q1\#B\#R$, where # is a special separator symbol. Thus, any Turing machine can be represented with a string encoding.

Example 14.1

The Turing machine for adding two unary numbers in Example 13.14 can be given by the following string:

3#q0#q1#q2#2#0#1#3#0#1#B#q0#B#1#q2#q0#0#q0#1#R#q0#1#q1#1#R#
q1#1#q1#1#R#q1#B#q2#B#R

Notice that all parts of the encoding use the # symbol as a separator. The initial 3 indicates that there are 3 states, which are then named right after. Next, the 2 indicates the size of Σ and these symbols then follow. The 3 indicates the size of Γ and these symbols follow. Next comes the start state q_0 followed by the blank symbol B. The size of F and its contents (just q_2) are given next. Finally, each of the δ transitions is given in the format explained above. For example, the first one q0#0#q0#1#R represents $\delta(q_0, 0) = (q_0, 1, R)$.

Given such an encoding of a Turing machine, it is possible (with a lot of effort!) to design a Turing machine that could take such an encoding and use it to process the behavior of the Turing machine on some input string. We could do this by, for example, storing the input string on one tape, the current state on another, the head position on a third, and the Turing machine encoding on a fourth. For each step of the Turing machine simulation, we would look up what character is at the current head position and what state we are currently at and then search the Turing machine encoding tape for the correct move to execute next.

The idea of the Universal Turing Machine is important for a number of reasons. For one, it tells us that Turing machines are capable of running other Turing machines. In fact, this is exactly what happens when you run a program using a compiler or interpreter—these are examples of Universal Turing Machine programs in that they can process arbitrary (correct) code.

Another important reason for Turing machines being able to simulate other Turing machines is that we can have a Turing machine call another Turing machine in a subroutine, allowing us to chain together increasingly complex computations.

14.5   RECURSIVE AND RECURSIVELY ENUMERABLE LANGUAGES

We are now ready to define what languages are computable.

> **Definition 14.2**
>
> A language is called **recursive** (or **computable** or **decidable**) if there exists a Turing machine that accepts the language and halts on all inputs.

Note that we specify that the Turing machine for the language must halt on all inputs. This is because Turing machines are allowed to reject a string by running forever. Since computations that run forever are not very useful, we specify that the Turing machine must halt and accept or halt and reject on every possible input.

There is also a broader collection of languages that defines everything a Turing machine is capable of computing called the recursively enumerable languages.

> **Definition 14.3**
>
> A language is called **recursively enumerable** if there exists a Turing machine that accepts the language.

Notice that this collection of languages contains the recursive ones because recursive languages all have Turing machines (that also must halt on all inputs). This classification isn't as useful from a practical standpoint because it allows Turing machines to run forever on some inputs, but it does help us with some proofs later in this chapter. Next, we prove a few results based on these definitions.

> **Theorem 14.1**
>
> Any language L is recursive if and only if $\overline{L}$ is recursive.

> **Proof**
>
> Let L be a recursive language. Then by the definition of a recursive language, L must have a Turing machine M that accepts it and halts on all inputs. We will design a new Turing machine M' that accepts $\overline{L}$ and halts on all inputs, thereby showing that $\overline{L}$ is recursive as well.
>
> Since M halts on all inputs, it must reject by reaching a state / symbol combination undefined for it. We thus define $M's$ δ function to be identical to that for M except for any missing transitions it should go to a new accepting state. Thus, all strings rejected by M would be accepted by M'. Secondly, the accepting state(s) for M are all made non-accepting for M', so that any computation that reaches them now halts and rejects any strings that were accepted by M. This newly constructed Turing machine M' now halts on all inputs and accepts precisely the language $\overline{L}$. It follows that $\overline{L}$ is recursive.
>
> The other direction follows from replacing L with $\overline{L}$ in the above proof. Thus, L is recursive if and only if $\overline{L}$ is recursive.

Theorem 14.2

The language L is recursive if and only if L and $\overline{L}$ are recursively enumerable.

Proof

Let L be any recursive language. Then, by definition, L has a Turing machine for it that halts on all inputs. Since L has a Turing machine for it, L is also recursively enumerable. By Theorem 14.1 we know that $\overline{L}$ is also recursive and thus also recursively enumerable. We have thus shown that if L is recursive, then L and $\overline{L}$ are both recursively enumerable.

Now we will assume that L and $\overline{L}$ are both recursively enumerable. Then, by definition, L has a Turing machine that accepts it, call it M, and $\overline{L}$ has a Turing machine that accepts it, call it $\overline{M}$. We can design a Turing machine M' for L that halts on all inputs as follows. On any input, M' will simulate M on the input and in parallel simulate $\overline{M}$ on another copy of the input (perhaps on a second tape), alternating steps for these two simulations. We know that we can have a Turing machine simulate another Turing machine based on the existence of the Universal Turing Machine. If the simulation of M ever halts and accepts, then M' will halt and accept. If the simulation of $\overline{M}$ ever halts and accepts, then M' will reject. Notice that M' accepts exactly the language L since M is a Turing machine for L and $\overline{M}$ is a Turing machine for $\overline{L}$. Moreover, exactly one of M and $\overline{M}$ must halt and accept because the input is either in L (in which case M will accept) or it is not (in which case $\overline{M}$ will accept). Since M' is a Turing machine that accepts L and halts on all inputs, this means that L must be recursive.

We have thus proved that L is recursive if and only if L and $\overline{L}$ are recursively enumerable.

Notice in the above proof that we had to run the simulations for M and $\overline{M}$ in parallel because it could be possible for either M or $\overline{M}$ to run forever on rejection and so if we had run them sequentially the computation may never end.

14.6   A NON-COMPUTABLE PROBLEM

To show that there are non-recursive problems, we first need to understand that all strings and Turing machines can be enumerated (listed in a numbered list). To list off all the strings in a fixed order, we can use shortlex ordering. For a review of shortlex ordering of strings, take a look at Section 1.10. To list all Turing machines, we can write an encoding for each one (see Section 14.4) and then also use a shortlex ordering on these. Let us call the Turing machines generated in this manner M_i for each string i.

We are now ready to define the Diagonal Problem (D). The Diagonal Problem can be intuitively thought of as the set of strings that when interpreted as a Turing machine accepts itself as an input.

Definition 14.4

$$D = \{i : M_i \text{ accepts input } i\}$$

To understand the above definition, we have to interpret the string i in two distinct ways. In one interpretation, we treat the string i as the encoding for a Turing machine M_i as we saw in Section 14.4. If the string i does not correspond to the encoding of any Turing machine, then we simply define M_i to be a simple Turing machine that rejects all inputs by immediately halting and rejecting. The second interpretation of i is simply that of a string—an input string to be accepted or rejected.

Next, we define the complement of the language D, called $\overline{D}$.

Definition 14.5

$$\overline{D} = \{i : M_i \text{ rejects input } i\}$$

We will now show that $\overline{D}$ does not have a Turing machine that accepts it. Before formally proving this, let us build some intuition about why this is the case. We use a technique called **diagonalization**.

Imagine that we could construct an infinitely large table that has all the Turing machines on the rows and the strings on the columns, as illustrated in Figure 14.3. For simplicity, we will assume that the strings are over the alphabet $\{0, 1\}$. Each entry in the table will indicate whether the Turing machine on the row accepts (A) the string on the column or rejects (R) it. (The entries in Figure 14.3 are not real and just shown for illustrative purposes.) We will focus on the diagonal entries of the table (marked with boxes) as the diagonal language and its complement are about these values. We can show that M_λ is not a Turing machine for $\overline{D}$ because (in this example table) λ is accepted by M_λ, putting λ in D (by the definition of D) but then a Turing machine for $\overline{D}$ should reject λ. A similar argument shows that M_0 is also not a valid Turing machine for $\overline{D}$. On the other hand, the Turing machine M_1 rejects the string 1, which means that 1 is in $\overline{D}$, but now M_1 cannot be a Turing machine for $\overline{D}$ as it should have accepted 1. Continuing this argument for each diagonal entry (whether it is A or R), shows that no Turing machine in our list of all possible Turing machines can accept $\overline{D}$. This shows that $\overline{D}$ is not recursively enumerable.

	λ	0	1	00	01	...
M_λ	A	R	R	R	R	...
M_0	A	A	A	A	A	...
M_1	R	R	R	R	R	...
M_{00}	R	A	A	R	A	...
M_{01}	A	A	R	R	A	...
...	...	...	...	...	...	...

Figure 14.3 Table of whether each Turing machine accepts/rejects each string

Theorem 14.3

The language $\overline{D}$ is not recursively enumerable.

Proof

We will prove that there cannot exist any Turing machine that accepts $\overline{D}$ via contradiction. Assume that $\overline{D}$ is accepted by a Turing machine with encoding i. We have two possible cases:

M_i accepts i: If it is the case that M_i accepts i, then by the definition of D it must be that $i \in D$. Then M_i cannot be a Turing machine for $\overline{D}$ since it should reject strings in D such as i.

M_i rejects i: If it is the case that M_i rejects i, then by the definition of $\overline{D}$ it must be that $i \in \overline{D}$. Then M_i cannot be a Turing machine for $\overline{D}$ since it should accept strings in $\overline{D}$ such as i.

We get a contradiction in both cases. Thus, a Turing machine for $\overline{D}$ cannot exist and so $\overline{D}$ is not recursively enumerable.

We can conclude from this the following result for D.

Theorem 14.4

The language D is not recursive.

Proof

We showed earlier in Theorem 14.2 that a language is recursive if and only if the language and its complement are recursively enumerable. Since we know that the complement of D is not recursively enumerable, it must be that D is not recursive.

We have shown that $\overline{D}$ and D are not recursive or computable, but they are not very natural languages. We will use them to next show that other, more easy to understand problems, are not computable as well.

14.7 REDUCTIONS

We will now show how to prove the non-computability of another language by bootstrapping from the previous two languages using a technique called a reduction. Consider the following language:

Definition 14.6

$A = \{(i, x) : M_i \text{ accepts } x\}$.

The language A consists of all pairs of Turing machine encodings and strings such that the Turing machine accepts the string. We encode pairings such as this by introducing new symbols (,), and ,. If we could compute this language, we could then tell whether any arbitrary program accepts an arbitrary input. Unfortunately, no Turing machine (that halts on all inputs) exists for it as we see next.

Theorem 14.5

The language A is not recursive.

Proof

We will assume that A is recursive for a contradiction. Then there exists a Turing machine M for A that halts on all inputs. We will use M to design a Turing machine for D that halts on all inputs.

We design a Turing machine M' that on input i will run M on the input (i, i) and then do what it does (accept if M accepts or reject if M rejects). We know that it is possible to design a Turing machine that runs another machine based on what we learned about the Universal Turing Machine.

We now show that M' is a Turing machine for D that halts on all inputs. This has three parts:

The Turing machine M' must halt on all inputs as the machine M halts on all inputs.

If $i \in D$, then M_i accepts i (by the definition of D), so M accepts (i, i) and thus M' accepts.

If $i \notin D$, then M_i rejects i (by the definition of D), so M rejects (i, i) and thus M' rejects.

The last two parts together give us that $i \in D$ if and only if M' accepts i, which means that M' correctly accepts all strings in D and rejects all others, making it a Turing machine for D that halts on all inputs. However, we already proved that D is not recursive, which gives us a contradiction. Thus, A is not recursive.

Intuitively, the above reduction uses the fact that determining whether M_i accepts an arbitrary string x should be even harder than determining whether M_i accepts the specific string i.

Since we believe Turing machines to be able to perform any mechanical computation possible on a computer, we can rephrase the implications of the above theorem thus:

> It is impossible to design a program that can predict the output of an arbitrary program on arbitrary input.

14.8   PROGRAM COMPARISON

We'll next see a reduction that is a little more involved. The notation $L(M_a)$ denotes the language of the Turing machine M_a.

Theorem 14.6

The language
$$SAME = \{(a, b) : L(M_a) = L(M_b)\}$$
is not recursive.

The language $SAME$ is conceptually the pairs of Turing machines that behave identically on all inputs. Translated from Turing machines to more practical usage, the uncomputability of $SAME$ implies the following:

> It is impossible to design a program that can tell if two programs behave identically on all inputs.

By showing that this problem is non-computable, this means that it is impossible for your CS instructor to design a program that takes as input two programs (for example, a working solution and a student's submission) and determine whether they behave identically on all inputs.

Proof

We will assume that $SAME$ is recursive for a contradiction. Then there exists a Turing machine M for $SAME$ that halts on all inputs. We will use M to design a Turing machine M' for D that halts on all inputs.

We design a Turing machine M' that on input i will create two Turing machines M_a and M_b as follows. The Turing machine M_a will ignore its input and instead just simulate M_i on input i doing whatever it does (accept if it accepts, reject if it rejects, or run forever if it runs forever). We know that it is possible to design a Turing machine that runs another machine based on what we learned about the Universal Turing Machine. The Turing machine M_b accepts all strings by making the start state accepting. Finally, the Turing machine M' runs the Turing machine M (for $SAME$) on input (a, b) and does whatever it does (accept or reject).

We now show that M' is a Turing machine for D that halts on all inputs. We prove the following three facts:

We know that M' halts on all inputs as the machine M halts on all inputs.

If $i \in D$, then M_i accepts i (by the definition of D) which means that M_a will always accept and hence $L(M_a) = \Sigma^*$. Moreover, $L(M_b) = \Sigma^*$ by design, so $L(M_a) = L(M_b)$ and thus M will accept, causing M' to accept.

If $i \notin D$, then M_i rejects i (by the definition of D) which means that M_a will always reject and hence $L(M_a) = \emptyset \neq \Sigma^* = L(M_b)$. Thus, M will reject, causing M' to reject.

Putting this together, gives us that $i \in D$ if and only if M' accepts, making M' a Turing machine for D that halts on all inputs. However, we already proved that D is not recursive, which gives us a contradiction. Thus, $SAME$ is not recursive.

At the heart of this reduction is the design of the Turing machine M_a that ignores its input and simulates M_i on input i. It is designed specifically so that for any $i \in D$

the Turing machine M_a will always accept giving $L(M_a) = \Sigma^*$ and for all $i \notin D$ the Turing machine M_a always rejects giving $L(M_a) = \emptyset$. This stark dichotomy allows us to use M_a in the reduction to distinguish whether a given string is in D or not by using a Turing machine for $SAME$.

More specifically, we have set up the Turing machines M_a and M_b that are given as input to the Turing machine for $SAME$ as follows:

$i \in D$?	M_i accepts i?	$L(M_a)$	$L(M_a) = L(M_b)$?	M accepts	M' accepts
Yes	Yes	Σ^*	Yes	Yes	Yes
No	No	$\emptyset$	No	No	No

The proof then takes the following form:

1. Assume for a contradiction that some Turing Machine M accepts $SAME$ and halts on all inputs.

2. Build a new machine M' that on input i halts and accepts if $i \in D$ and halts and rejects if $i \notin D$. It does so by running M (the Turing machine for $SAME$) on the Turing machines M_a and M_b described above. (This is the part of the proof that you will have to adapt, using a table similar to the one above, to solve most of the exercises.) Since M halts on all inputs, so must M'.

3. Since we showed in the last step that M' is a Turing machine for D that halts on all inputs, this shows that D is recursive. This is a contradiction, so $SAME$ can not be recursive.

14.9   THE HALTING PROBLEM

Another classic problem that is known to be uncomputable is the **Halting Problem**. Simply put, the Halting Problem asks whether a given computation will ever halt or if it will run forever. By showing that this problem is not recursive we know, for example, that it is impossible to design a compiler that can always detect infinite loop bugs in arbitrary code.

Next, we define the Halting Problem and show that it is undecidable as well.

Definition 14.7

$H = \{(i, x) : M_i \text{ halts on input } x\}$.

Notice that this definition is about whether a Turing machine halts or not on a given input, making it about the behavior of the Turing machine beyond whether it accepts certain strings or not. This makes the reduction a little more complicated.

Theorem 14.7

The language H is not recursive.

Proof

We will assume that H is recursive for a contradiction. Then there must exist a Turing machine M for H that halts on all inputs. We will use M to design a Turing machine for D that halts on all inputs.

We design a Turing machine M' that on input i will compute the encoding i' of a Turing machine $M_{i'}$ that will behave the same as M_i, except if M_i is about to halt and reject $M_{i'}$ will go into an infinite loop. Next, M' will run the Turing machine M on the input (i', i). We can imagine designing a Turing machine that automatically modifies another Turing machine to behave in the above manner since it is easy to add a state for going into an infinite loop by, for example, moving right on the tape forever. We also know that it is possible to design a Turing machine that runs another machine based on what we learned about the Universal Turing Machine.

We now show that M' is a Turing machine for D that halts on all inputs. We show the following three parts:

The Turing machine M' must halt on all inputs as the machine M halts on all inputs.

If $i \in D$, then M_i accepts i (by the definition of D), so $M_{i'}$ halts on input i, and thus it must be the case that (i', i) is accepted by M, so M' accepts.

If $i \notin D$, then M_i rejects i (by the definition of D), so by design $M_{i'}$ will reject i by running forever, and thus (i', i) is rejected by M, hence M' rejects.

The last two parts together give us that $i \in D$ if and only if M' accepts, making M' a Turing machine for D that halts on all inputs. However, we already proved that D is not recursive, which gives us a contradiction.

Thus, H is not recursive.

In practical terms:

> It is impossible to design a program that can tell if another program will ever terminate.

The Halting Problem is one of the classic problems in computer science that is uncomputable.

14.10 CLASSES OF LANGUAGES

When studying the difficulty of solving a problem, computer scientists use the notion of a class of problems.

Definition 14.8

A **class** of problem is a set of languages that share a common characteristic in terms of how hard it is to compute them.

Example 14.2

> Some examples of classes of problems that we have studied in this book include the regular languages (Chapter 2), the context-free languages (Chapter 12), the recursive and recursively enumerable languages (Chapter 14).

In conclusion, here is a diagram with the containment properties of these classes.

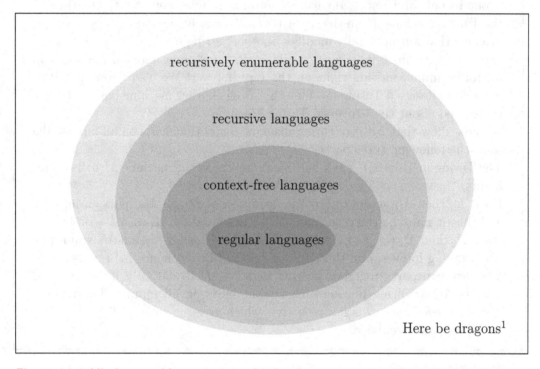

Figure 14.4 All classes of languages in this book

14.11 CHAPTER SUMMARY AND KEY CONCEPTS

- **Variations** of Turing-machine models include ones with **multiple tracks** and **multiple tapes**. These models are no more powerful than the standard Turing machine model.

- The **Church-Turing thesis** states that all mechanical computation can be performed by a Turing machine.

- The **Universal Turing machine** is one that takes as input the encoding of a Turing machine and an input and simulates that Turing machine on that input.

- A language is called **recursive** or **computable** or **decidable** if there exists a Turing machine that accepts it and halts on all inputs.

---

[1] You have learned that there are problems unsolvable by any computer—dragons!—hence the cover of this book.

- A language is called **recursively enumerable** if there exists a Turing machine that accepts it.

- A language is recursive if and only only if its complement is recursive.

- A language is recursive if and only if it and its complement are recursively enumerable.

- The **diagonal language** D and its complement can be shown to be non-recursive using **diagonalization**.

- **Reductions** are a way of showing that a language is not recursive. To do this, you must reduce from a problem known not to be recursive to the problem being shown is not recursive.

- Many problems, such as determining (1) whether a Turing machine (program) will accept a given input, (2) whether two Turing machines (programs) accept the same language, and (3) whether a Turing machine (program) will ever halt are all known to be non-recursive.

- A **class** of problem is a set of languages that have the same computational difficulty. These include **regular**, **context-free**, **recursive**, and **recursively enumerable** languages.

EXERCISES

Read the entries in the Stanford Encyclopedia of Philosophy on Computability (https://plato.stanford.edu/entries/computability/) and Alan Turing (https://plato.stanford.edu/entries/turing/) and other sources to give short answers to the following questions:

14.1 Who designed the Lambda Calculus?

14.2 What was Gödel's Incompleteness Theorem?

14.3 What was the *entscheidungsproblem*?

14.4 What was the title of the paper in which Turing introduced the Turing machine?

14.5 What is the Turing Test?

14.6 Do you think that a Turing machine can simulate a human brain?

Prove that the following problems are undecidable using reductions. For each problem, write a sentence explaining in words what the implication of the undecidability of the language is for computer programs.

14.7 $ALL = \{i : L(M_i) = \Sigma^*\}$

14.8 $INFINITE = \{i : L(M_i) \text{ is infinite}\}$

14.9 $NONEMPTY = \{i : L(M_i) \neq \emptyset\}$

14.10 $HASEMPTY = \{i : \lambda \in L(M_i)\}$

14.11 $FULLCOVER = \{(i,j) : L(M_i) \cup L(M_j) = \Sigma^*\}$

14.12 $SUBSET = \{(i,j) : L(M_i) \subseteq L(M_j)\}$

14.13 $DIFFERENT = \{(i,j) : L(M_i) \neq L(M_j)\}$

14.14 $OVERLAP = \{(i,j) : L(M_i) \cap L(M_j) \neq \emptyset\}$

14.15 $EMPTY = \{i : L(M_i) = \emptyset\}$ (Hint: Reduce from $\overline{D}$ rather than D)

14.16 $DISJOINT = \{(i,j) : L(M_i) \cap L(M_j) = \emptyset\}$ (Hint: Reduce from $\overline{D}$)

14.17 (Harder) $FINITE = \{i : L(M_i) \text{ is finite}\}$

14.18 (Harder) $SIZE5 = \{i : L(M_i) \text{ has exactly 5 strings}\}$

14.19 (Harder) $ONESTAR = \{i : L(M_i) = 1^*\}$

14.20 (Harder) $REGULAR = \{i : L(M_i) \text{ is regular}\}$

Counting

A.1 WHY YOU SHOULD CARE

Being able to count up the number of different ways or combinations that are possible for a problem is an exceedingly useful skill in computer science. It is needed for tasks as different as counting the number of passwords (to improve computer security), determining the number of ways processes can be scheduled, to measuring the search space for a variety of games so that you can compute an optimal play strategy.

A.2 THE MULTIPLICATION RULE

Remember back in Chapter 3 when we created truth tables? A truth table with two variables had 4 rows but another with three variables had 8 rows. If we added another variable, we would need a total of 16 rows. The number of rows doubles each time we add a variable. Why is this?

The easiest way to think about the number of rows needed by a truth table is to consider the possibilities. The first variable can be T or F for a total of 2 possibilities. When we add a second variable, each of the original possibilities is now paired with a T or F for the second variable for a total of $2 \times 2 = 4$. Continuing in this manner, if we have n variables, we need $2 \times 2 \times \ldots \times 2$ (n times) or 2^n rows. Put another way, every time we add a variable we need to have two copies of the previous table: one for when the new variable is T and another for when it is F. This is a demonstration of the multiplication rule.

> **Definition A.1**
>
> If there are c_1 possibilities for a first option, c_2 possibilities for a second option, $\ldots$, c_n possibilities for an nth option, then the total number of combinations of all the options is the product of the terms $c_1 c_2 \ldots c_n$.

It is important to note here that repeated values are allowed, e.g., more than one variable can be T in the same row of a truth table. The count is quite different when values are not allowed to be repeated and this will be explored further in later sections of this chapter.

DOI: 10.1201/9781003383284-A

Example A.1

The number of distinct values that can be represented by an n-bit number can be computed by considering that there are two possibilities for each bit (0 or 1) so that $c_1 = c_2 = \ldots c_n = 2$, giving a total number of $2 \times 2 \times \ldots \times 2$ (n times) or 2^n distinct values.

Since the number with all 0s represents the number zero, this means that an n-bit counter can count up to $2^n - 1$. For example, in many programming languages that have 32-bit integers, arithmetic will overflow when doing computation on values larger than $2^{32} - 1$ (about 4 billion) when using unsigned integers or $2^{31} - 1$ (about 2 billion) when using signed integers.

Example A.2

Devices on the Internet used to be assigned a unique IP address (IPv4) that was represented by a dotted quad such as 140.141.2.0. Each of the four numbers in an IP address is an integer between 0 and 255 (each of the four numbers is represented by 8 bits and $2^8 - 1 = 255$). Since there are $c_1 = c_2 = c_3 = c_4 = 256$ options for each of the four numbers, the total number of IPv4 addresses is 256^4 or about 4 billion. Since this was clearly insufficient for the growing number of devices on the Internet, the most recent version (IPv6) uses 128-bits (rather than the 32-bits of IPv4) to give 2^{128} or about 3.4×10^{38} unique addresses.

Example A.3

If you use a password manager, it will generate random passwords for you that are very hard to be guessed using an exhaustive search. If uppercase and lowercase letters and the ten digits are used by the password manager, then there are a total of 62 possibilities for each character. A randomly generated password that is twenty characters long can then be any of 62^{20} or about 7×10^{35} different options. Even trying a million passwords a second, it would take over 2×10^{22} years to guess such a password. In contrast, a shorter password (still random) of length 8 would be one of 62^8 or about 2×10^{14} options and at the same rate would take less than seven years to be guessed.

A.3 ARRANGEMENTS WITHOUT REPEATS, ORDER MATTERS

Let us enumerate the number of ways you can rearrange a list of three items (say, 1, 2, and 3). The different ways are $[1, 2, 3], [1, 3, 2], [2, 1, 3], [2, 3, 1], [3, 1, 2], [3, 2, 1]$, for a total of six arrangements. Where did this number six come from? Well, we had three options for the first number (1, 2, or 3), followed by any of the two remaining numbers for the second, followed by a single option for the third. Multiplying these out, we get $3 \times 2 \times 1$ or 6.

Extending this idea, we get that the number of ways to arrange a list of size n is $n \times (n-1) \times \ldots \times 1$. We give a special name to this function called factorial.

Definition A.2

The factorial of an integer n, written as $n!$, is the product $1 \times 2 \times 3 \times \ldots \times n$. By definition, we take $0!$ to be 1.

Example A.4

The number of ways to rearrange a list of ten distinct items is $10!$ or about 3.6 million. The number of ways to rearrange a list of a hundred distinct items is $100!$ or about 9×10^{157}. Incredibly, any sorting algorithm can correctly rearrange all of these arrangements into sorted order—an impressive feat!

We can think of rearrangements without repetitions as a special case of the multiplication rule in which $c_1 = c_2 + 1 = c_3 + 2 = \ldots$.

Example A.5

The number of ways to draw three cards (in order) from a standard 52 card deck of playing cards is $52 \times 51 \times 50$.

A.4 ARRANGEMENTS WITHOUT REPEATS, ORDER DOESN'T MATTER

When counting the number of hands in a card game the order in which you get dealt cards usually doesn't matter. For example, getting the 2, 3, and 4 of spades is the same as getting the 4, 3, and 2 of spades. In such situations, where the order in which items appear in the arrangement are immaterial, we are effectively over-counting the number of arrangements by the number of arrangements of that size. For instance, when drawing three cards, there are $3!$ or 6 different ways to be dealt the same set of three cards. To correct for this over-counting, we can divide the total number of rearrangements by six.

Example A.6

The number of ways to draw three distinct cards from a standard 52 card deck of playing cards is

$$(52 \times 51 \times 50)/(1 \times 2 \times 3).$$

The number of ways to be dealt a five-card poker hand is

$$(52 \times 51 \times 50 \times 49 \times 48)/(1 \times 2 \times 3 \times 4 \times 5).$$

In general, if you are picking exactly k items without repetition from a collection of n items and the order doesn't matter, then we use combinations.

Definition A.3

The number of ways to choose exactly k items without repetition from among n choices is given by $\binom{n}{k}$ (pronounced "n choose k") and is given by the formula:

$$\binom{n}{k} = \frac{n \times (n-1) \times \ldots \times (n-k+1)}{1 \times 2 \times \ldots k} = \frac{n!}{(n-k)!k!}.$$

Example A.7

The number of ways to choose a group of 3 students from a class of size 24 is

$$\binom{24}{3} = \frac{24!}{21!3!} = \frac{24 \times 23 \times 22}{1 \times 2 \times 3} = 2024.$$

A.5   CHAPTER SUMMARY AND KEY CONCEPTS

- The **multiplication rule** is that if there are c_1 possibilities for a first option, c_2 possibilities for a second option, ..., c_n possibilities for an nth option, then the total number of combinations of all the options is $c_1 c_2 \ldots c_n$.

- The number of arrangements of n distinct items when there are **no repeats and order matters** is $n!$.

- The number of arrangements of k items out of n distinct items when there are **no repeats and order doesn't matter** is $\binom{n}{k} = \frac{n!}{(n-k)!k!}$.

EXERCISES

Multiplication rule (repeats allowed, order matters)

A.1 How many license plates are possible with a three letter combination followed by a 4 digit number? (Note: the number may have a leading 0.)

A.2 How many variable names of length 5 are possible? Recall that a variable name must start with a letter (upper or lower case) or underscore character and the other positions can have a letter (upper or lower case), underscore, or digit.

A.3 Suppose you are required to enter an eight character password. The first character must be a letter (upper or lower case). The next 6 characters can be upper/lower case letters, or digits 0-9. The last character must be one of these punctuation marks: underscore, asterisk, exclamation point, caret, dollar sign, percent sign, or ampersand. How many different passwords are possible?

A.4 How many passwords are possible when you string together four random English words (repetitions allowed)? You may assume that there are 170,000 words in the English language. (Relevant XKCD: https://xkcd.com/936/)

Permutations (no repeats, order matters)

A.5 Four players are engaged in a Scrabble competition. There is a large prize for first place and a smaller one for second place. How many different ways can these two prizes be distributed?

A.6 How many different ways can you select and arrange four out of ten cards (in order) in your hand?

A.7 Three friends are have four ice pops to choose from: apple, mango, orange, and grape. In how many ways can they pick one each?

A.8 There are nine new movies available on your streaming service. In how many ways can you watch one different movie each night this week?

A.9 You're going on a road trip with seven other friends and have room for three of them in your car. If it matters which seat each person takes (front passenger's seat, rear left, and rear right), then how many ways are there to pick friends for your car?

Basic combinations (no repeats, order doesn't matter)

A.10 How many ways are there to be dealt a seven-card hand from a standard 52 card deck?

A.11 How many ways are there to select a four person group from a class of 20 students?

A.12 How many different sums of money can you make from exactly two coins out of a penny (1 cent), a nickel (5 cents), a dime (10 cents), and a quarter (25 cents)?

A.13 Suppose that 20 people all decide to high five each other. How many high fives would this amount to?

A.14 Ten kids are making teams to play soccer against one another. First, two captains are selected. The captains then alternate picking kids for their team until all the kids are on a team. How many different ways are there to end up with a pair of teams selected in this manner?

A.15 A lottery involves selecting six numbers without replacement from 1 to 70. If you select six of these numbers at random, what is the probability of winning?

Probability

B.1 WHY YOU SHOULD CARE

Computer scientists need to know about probability to be able to design randomized algorithms—ones that have random choices made in them. Randomized algorithms are useful because they can either improve the expected running time (these are called Las Vegas algorithms) or allow faster computation at the expense of a small probability of failure (called Monte Carlo algorithms). In addition, randomness is important for designing data structures such as hash functions, computing in scientific applications such as Monte Carlo simulations, and making video game AI opponents less predictable.

B.2 SAMPLE SPACES

When studying probabilities of events, we must first consider the sample space of possibilities. For example, when you roll a six-sided die there are six possible outcomes or possible rolls. When you flip a coin there are two outcomes (heads or tails). This set is called the sample space.

Definition B.1

The **sample space** of a random trial is the set of possible outcomes for that trial. The probability of an outcome x is denoted by $\Pr(x)$, where $0 \leq \Pr(x) \leq 1$. Moreover, the sum of the probabilities of all the outcomes in the sample space must equal 1.

Example B.1

When flipping a coin, the sample space is $\{heads, tails\}$. If the coin is a fair one, then

$$\Pr(\text{flip} = heads) = 1/2$$

and

$$\Pr(\text{flip} = tails) = 1/2.$$

DOI: 10.1201/9781003383284-B

Example B.2

When rolling a six-sided die, the sample space is $\{1, 2, 3, 4, 5, 6\}$. Since the probability of each of these is equal, we can write

$$\Pr(\text{roll} = 1) = 1/6, \Pr(\text{roll} = 2) = 1/6, \ldots, \Pr(\text{roll} = 6) = 1/6.$$

B.3 EVENTS

When studying probability, we will often want to group together a number of outcomes and refer to them collectively as an event.

Definition B.2

Any subset of a sample space S is called an **event**.

The probability of an event is simply the sum of probabilities of the outcomes that make up that event.

Example B.3

The probability of rolling an even number with a six-sided die is

$$
\begin{aligned}
\Pr(\text{roll is even}) &= \Pr(\text{roll} = 2) + \Pr(\text{roll} = 4) + \Pr(\text{roll} = 6) \\
&= 1/6 + 1/6 + 1/6 \\
&= 1/2.
\end{aligned}
$$

Example B.4

The probability of rolling less than 3 with a six-sided die is

$$
\begin{aligned}
\Pr(\text{roll less than 3}) &= \Pr(\text{roll} = 1) + \Pr(\text{roll} = 2) \\
&= 1/6 + 1/6 \\
&= 1/3.
\end{aligned}
$$

Definition B.3

The **complement** of an event is the complement of the event set with the sample space being the universal set.

Example B.5

The complement of rolling an even number (i.e., the event $\{2, 4, 6\}$) is rolling an odd number (i.e., the event $\overline{\{2, 4, 6\}} = \{1, 3, 5\}$).

The probability of the complement of an event is one minus the probability of the event itself.

The complement of rolling less than 3 is not rolling less than a 3 (i.e., rolling 3 or higher). This can be computed as

$$\begin{aligned} \mathrm{Pr}(\text{roll not less than 3}) &= 1 - \mathrm{Pr}(\text{roll less than 3}) \\ &= 1 - 1/3 \\ &= 2/3. \end{aligned}$$

Definition B.4

A pair of events is called **mutually exclusive** if they are non-overlapping subsets of the sample space.

For mutually exclusive events, the probability that either event happens is the sum of the probabilities of the two events.

Example B.7

The event of getting a value less than 3 on a die roll is mutually exclusive from getting larger than a 4 since the subset of the sample space is $\{1, 2\}$ in the former case and $\{5, 6\}$ in the latter and these sets have no common elements. Thus, the probability of getting a value that is either less than 3 or more than 4 is $\mathrm{Pr}(\text{roll less than 3}) + \mathrm{Pr}(\text{roll more than 4}) = 2/6 + 2/6 = 2/3$.

A pair of events is called **independent** if neither event affects the probability of the other.

Example B.8

The rolling of two dice is independent as the outcome for either die doesn't affect the other. On the other hand, the event that we get heads on a coin flip is not independent from the event of getting tails on the same coin flip because either one happening makes the probability of the other zero.

We now recast the definition mathematically.

Definition B.5

We say that two events A and B are **independent** if and only if

$$\mathrm{Pr}(A \text{ and } B) = \mathrm{Pr}(A) \cdot \mathrm{Pr}(B).$$

Example B.9

The probability of getting a pair of sixes when rolling a pair of dice can be computed as

Pr(the first roll is a 6) · Pr(the second roll is a 6) = $(1/6) \cdot (1/6) = 1/36$.

Example B.10

The probability of *either* roll of two dice coming up six can be computed as the complement of the event that neither of them comes up six:

$$
\begin{aligned}
\text{Pr(either roll is 6)} &= 1 - \text{Pr(neither roll is 6)} \\
&= 1 - \text{Pr(first roll is not 6)} \cdot \text{Pr(second roll is not 6)} \\
&= 1 - (5/6) \cdot (5/6) \\
&= 11/36.
\end{aligned}
$$

B.4 CHAPTER SUMMARY AND KEY CONCEPTS

- The **sample space** of a random trial is the set of all possible outcomes of the trial.

- The **probability** of an outcome x is denoted by $\text{Pr}(x)$, where $0 \le \text{Pr}(x) \le 1$. The sum of the probabilities of all the outcomes in the sample space must be 1.

- An **event** is a subset of a sample space. The probability of the event is the sum of the probabilities of the outcomes that make up the event.

- The **complement** of an event is the complement of the event set.

- Two events are called **mutually exclusive** if their sets do not intersect.

- Two events are called **independent** if neither affects the probability of the other.

EXERCISES

A standard deck of cards has one each of the values Ace, 2 through 10, Jack, Queen, and King of each the suits Hearts (red ♥), Diamonds (red ♦), Clubs (black ♣), and Spades (black ♠) for a total of $13 \times 4 = 52$ cards.

B.1 What is the probability of drawing an Ace of Hearts from a deck?

B.2 What is the probability of drawing an Ace from a deck?

B.3 What is the probability of drawing a Diamond from a deck?

B.4 What is the probability of drawing an Ace or a King from a deck?

B.5 What is the probability of drawing a red card (Heart or Diamond) from a deck?

B.6 What is the probability of drawing an Ace or a Black card (Club or Spade) from a deck?

B.7 What is the probability of not drawing an Ace of Spades from a deck?

B.8 What is the probability of not drawing a Queen from a deck?

B.9 What is the probability of not drawing a King or Queen from a deck?

B.10 What is the probability of not drawing an Ace or a red card (Heart or Diamond) from a deck?

For the following questions, you should assume that cards are returned to the deck and shuffled before the next card is drawn (this is called "sampling with replacement"):

B.11 What is the probability of drawing an Ace and then a Queen from a deck?

B.12 What is the probability of drawing two Aces from a deck?

B.13 What is the probability of drawing an Ace and then a Black card (Club or Spade) from a deck?

B.14 What is the probability of drawing an Ace and then a card that's not an Ace from a deck?

B.15 What is the probability of drawing two cards neither of which is an Ace from a deck?

For the following questions, you should assume that cards are *not* returned to the deck before the next card is drawn (this is called "sampling without replacement"):

B.16 What is the probability of drawing an Ace and then a Queen from a deck?

B.17 What is the probability of drawing two Aces from a deck?

B.18 What is the probability of drawing an Ace and then a Black card (Club or Spade) from a deck?

B.19 What is the probability of drawing an Ace and then a card that's not an Ace from a deck?

B.20 What is the probability of drawing two cards neither of which is an Ace from a deck?

The following questions assume that you have already read Appendix A. In the game poker, each player is dealt a hand with five cards the order of which does not matter.

B.21 In poker, what is the probability of being dealt a royal flush (10, Jack, Queen, King, and Ace all of the same suit)?

B.22 In poker, what is the probability of being dealt a flush (all five cards of the same suit)?

B.23 In poker, what is the probability of being dealt a straight (consecutive cards such as Ace, 2, 3, 4, 5 or 10, Jack, Queen, King, Ace)?

B.24 In poker, what is the probability of being dealt a straight flush (consecutive cards of the same suit)?

B.25 In poker, what is the probability of being dealt a four-of-a-kind (four cards with the same value and one other)?

B.26 In poker, what is the probability of being dealt a full house (three cards of the same value and two other cards with the same value)?

B.27 In poker, what is the probability of being dealt a three-of-a-kind (three cards of the same value and the remaining two different from the rest and each other)?

B.28 In poker, what is the probability of being dealt a pair (two cards of the same value and all the remaining three different from the rest and each other)?

B.29 In poker, what is the probability of being dealt a two pair (two pairs of cards of the same value and a fifth card with value different from the rest)?

B.30 (Harder) In poker, what is the probability of not getting any of the above hands?

APPENDIX C

Elementary Number Theory

C.1   WHY YOU SHOULD CARE

Number theory is the basis of a number of algorithms involving hash functions, pseudo-random number generators, and cryptography. Most modern cryptography and cryptocurrencies would not be possible without it.

C.2   MODULAR ARITHMETIC

When we divide integers in Python, we are actually performing floor division. **Floor division** is when we divide an integer by a non-zero integer and round down.

Example C.1

For example, when we floor divide 7 by 2 this is written mathematically as $\lfloor 7/2 \rfloor$ (analogous to 7//2 in Python) and is equal to 3. Similarly, floor dividing 10 by 7 is $\lfloor 10/7 \rfloor$ which is equal to 1.

For negative numbers we also round down (smaller) so $\lfloor -7/2 \rfloor$ is equal to -4. Similarly, $\lfloor 10/-7 \rfloor$ is -2.

Some languages, such as C, C++, and Java, behave differently with regards to integer division with negative numbers. They round toward zero by discarding the fractional part of the division. This difference between language semantics is a common source of errors that you should try to avoid.

We can now define the quotient and remainder for division as follows:

Definition C.1

For any positive integers $a, b > 0$, we can write $a = qb + r$, where $q = \lfloor a/b \rfloor$ is called the **quotient** and $0 \leq r < b$ is called the **remainder**.

Example C.2

When we divide 25 by 10 we get a quotient of $\lfloor 25/10 \rfloor = 2$ and a remainder of 5. Note that $25 = 10 \cdot 2 + 5$.

DOI: 10.1201/9781003383284-C

The quotient and remainder can be computed using the following division algorithm: We get the remainder using an operation called the **modulus**, abbreviated to

---

Algorithm C.1: division(a, b)

input : Integers $a, b \geq 1$
output: The quotient and remainder when dividing a by b
1: $r = a$
2: $q = 0$
3: **while** $r \geq b$ **do**
4:     $r = r - b$
5:     $q = q + 1$
6: **return** (q, r)

---

mod. This is the same as the % operation in most programming languages, though it operates differently for negative numbers in different languages (as with division).

Example C.3

We write the remainder of when we divide 10 by 3 as 10 mod 3, which is equal to 1. Similarly, 25 mod 10 is 5.

The mod terminology is also sometimes used to show that two numbers are related as they have the same remainder when divided by the same divisor. This is called congruence modulo the divisor and written $a \equiv b \pmod c$ ("a is congruent to b modulo the divisor c").

Example C.4

We say that 10 and 13 are both congruent modulo 3 since 10 mod 3 = 1 = 13 mod 3. This is written mathematically as $10 \equiv 13 \pmod 3$.

We say that an integer a is divisible by another non-zero integer b (written $b \mid a$) if b evenly divides into a with no remainder. We say b is a **divisor** or **factor** of a.

Definition C.2

For any integers $a, b > 0$, the following are equivalent:

1. a is divisible by b ($b \mid a$),

2. a is a multiple of b,

3. b is a divisor of a,

4. b is a factor of a, and

5. a mod $b = 0$.

Example C.5

Since 4 evenly divides into 12, we can say that

1. 12 is divisible by 4 (4 | 12),

2. 12 is a multiple of 4,

3. 4 is a divisor of 12,

4. 4 is a factor of 12, and

5. 12 mod 4 = 0.

For a variety of numerical algorithms, such as reducing fractions to lowest terms, we need to compute the greatest common divisor (also called the highest common factor in some parts of the world) between a pair of numbers.

Definition C.3

The **greatest common divisor** of a pair of positive integers is the largest integer that divides both integers.

Example C.6

The greatest common divisor of 12 and 16 is $\gcd(12, 16) = 4$ since 4 divides both 12 and 16 and there is no larger integer that does so. Similarly, the greatest common divisor of 14 and 15 is $\gcd(14, 15) = 1$ since no integer greater than 1 divides both 14 and 15.

Another useful function on a pair of positive integers is the least common multiple that is used for finding the common denominator of two fractions.

Definition C.4

The **least common multiple** of a pair of positive integers is the smallest positive integer that is a multiple of both integers.

Example C.7

The least common multiple of 12 and 16 is $\mathrm{lcm}(12, 16) = 48$ since 48 is the smallest multiple of both 12 and 16. Similarly, the least common multiple of 14 and 15 is $\mathrm{lcm}(14, 15) = 210$.

The greatest common divisor and lowest common multiple are related to one another. For any positive integers a and b, we have that $\gcd(a, b) \cdot \mathrm{lcm}(a, b) = a \cdot b$. This means that we can easily compute one if we have the other. We usually compute the gcd using Euclid's algorithm that you will learn about next.

C.3 EUCLID'S ALGORITHM FOR GCD

Rather than try every possible divisor of a pair of numbers, Euclid's algorithm computes the gcd very efficiently and elegantly.

Algorithm C.2: Euclid(a, b)

---

input : Integers $1 \leq a \leq b$

output: The gcd of a and b

  1: **if** $b \bmod a = 0$ **then**

  2:     **return** a

  3: **else**

  4:     **return** Euclid$(b \bmod a, a)$

Example C.8

To compute the gcd of 16 and 28 (note that the algorithm assumes that the smaller number is given first), Euclid's algorithm proceeds as follows:

$$
\begin{aligned}
\text{Euclid}(16, 28) \quad &= \quad \text{Euclid}(28 \bmod 16, 16) = \text{Euclid}(12, 16) \\
&= \quad \text{Euclid}(16 \bmod 12, 12) = \text{Euclid}(4, 12) \\
&= \quad 4 \qquad\qquad\qquad (\text{as } 12 \bmod 4 = 0).
\end{aligned}
$$

C.4 CHAPTER SUMMARY AND KEY CONCEPTS

- **Floor division** is when integers are divided with no remainder.

- When dividing two positive integers, the result of the floor division is called the **quotient**. Any remaining amount is called the **remainder** and is computed using the **mod** operation.

- The **greatest common divisor** of a pair of positive integers is the largest integer that divides both.

- The **least common multiple** of a pair of positive integers is the smallest positive integer that is a multiple of both.

- **Euclid's algorithm** allows us to quickly compute the greatest common denominator for two positive integers.

EXERCISES

Compute the quotient and remainder in each of the following cases:

C.1 10 divided by 4

C.2 24 divided by 7

C.3 -12 divided by 5

C.4 17 divided by -10

Compute the gcd and lcm of the following pairs of numbers. Show the steps of Euclid's algorithm for the gcd.

C.5 21 and 63

C.6 10 and 16

C.7 15 and 23

C.8 17 and 64

C.9 19 and 128

C.10 128 and 144

Asymptotic Notation

D.1 WHY YOU SHOULD CARE

When analyzing the efficiency of our algorithms, computer scientists use asymptotic notation to capture such metrics as running time, memory used, number of queries, energy usage, etc. Analyzing the rate of growth of each of these quantities is far more important than computing the exact value for a few instances.

D.2 WHY ASYMPTOTIC NOTATION

Suppose that you are comparing a pair of algorithms and trying to decide which one you will implement for an application that you are working on. If you are interested in understanding which one will run quicker, you may count up the number of basic operations of each and use this for comparison. To make things concrete, say that you have one algorithm that, on an input size of n, needs $f(n) = 1000n$ operations and another that needs $g(n) = n^2$ operations. If you were to run these on small instances of your problem, you would likely find that the second algorithm runs faster. However, in practice computer scientists are wary of small instances as we know that data sizes are constantly increasing and we should instead plan ahead for large data sets. This is where asymptotic notation allows us to compare the above algorithms for larger sizes.

In the above example, we would say that the first algorithm is a $\Theta(n)$ (pronounced "theta of n") time algorithm and the second is $\Theta(n^2)$ time. In other words, the rate of growth of the second function is considerably higher than the first. For example, if we increase the size a thousand-fold, then the first algorithm's run-time goes up by a factor of a thousand but the second's run-time goes up by a factor of one million! When working with large size inputs the second algorithm will be several orders of magnitude slower than the first.

When performing asymptotic analysis, we ignore constant multiplicative factors such as the 1000 in $f(n) = 1000n$. In most algorithms these factors are not very large and don't significantly affect the analysis. We also ignore how algorithms perform on small input sizes as these are not as relevant to us most of the time. Both these considerations have lead to computer scientists using asymptotic notation as the way we analyze the performance of our algorithms.

DOI: 10.1201/9781003383284-D

D.3 THETA NOTATION

The most useful form of asymptotic notation is theta notation since it gives an exact characterization of the growth rate of the function. When in doubt about which type of asymptotic notation you should be using, theta notation is your best bet since it is the most precise.

Definition D.1

We say that $f(n)$ is $\Theta(g(n))$ if there exists $n_0, c_1, c_2 > 0$ such that for all $n \geq n_0$ we have that $c_1 g(n) \leq f(n) \leq c_2 g(n)$.

In other words, if we can sandwich the function $f(n)$ between constant multiples of $g(n)$ for sufficiently large n, then we can say that $f(n) = \Theta(g(n))$.

Example D.1

The function $f(n) = 1000n$ is $\Theta(n)$ with constants $n_0 = 13, c_1 = 500, c_2 = 1001$ because, for all $n \geq 13$, $500n \leq f(n) \leq 1001n$.

Notice that there could be several different constants n_0, c_1, c_2 that could also work. Which ones you pick is not important as long as they show the above sandwich property.

Example D.2

The function $g(n) = n^2$ is $\Theta(n^2)$ with constants $n_0 = 1, c_1 = 1/2, c_2 = 1$ because, for all $n \geq 1$, $n^2/2 \leq g(n) \leq n^2$.

D.4 BIG-O AND BIG-Ω NOTATION

Sometimes we are unable to, or don't care to, compute the exact characterization of a function and are happy to settle for an upper or lower bound. For example, we may want to argue that our sorting routine takes at most quadratic time. In such a case, we might use Big-O for an upper bound and Big-Ω ("big omega") for a lower bound.

Definition D.2

We say that $f(n)$ is $O(g(n))$ if there exists $n_0, c > 0$ such that for all $n \geq n_0$ we have that $f(n) \leq cg(n)$.

This is identical to the definition and examples introduced in Section 3.7. Since Big-O notation is only an upper bound, it may not be a precise bound on the function that you are studying. This is why you should use theta notation if you are trying to be precise. Big-O notation can give an overestimate of the growth rate of a function as can be seen in the following example:

Example D.3

The function $f(n) = 1000n$ is $O(n^2)$ with constants $n_0 = 1, c = 1000$ because, for all $n \geq 1$, $f(n) \leq 1000n^2$.

Similar to upper bounds, we sometimes have need for lower bounds on the growth of a function. For example, we may want to argue that the runtime of an algorithm is always at least quadratic.

Definition D.3

We say that $f(n)$ is $\Omega(g(n))$ if there exists $n_0, c > 0$ such that for all $n \geq n_0$ we have that $cg(n) \leq f(n)$.

Example D.4

The function $f(n) = 5n^2$ is $\Omega(n)$ with constants $n_0 = 1, c = 5$ because, for all $n \geq 1$, $f(n) \geq 5n$.

Once again note that since Big-Ω notation is only a lower bound, it may not be a tight bound.

D.5   STRICT BOUNDS

All the above bounds were inclusive ones that included the function being compared with. In some circumstances we want to say that the asymptotic growth rate of a function is strictly smaller than, or strictly larger than, another function. This is where little-o and little-ω ("little omega") notation is used.

Definition D.4

We say that $f(n)$ is $o(g(n))$ if for all $c > 0$ there exists a constant $n_0 > 0$ such that for all $n \geq n_0$ we have that $f(n) < cg(n)$.

Example D.5

The function $f(n) = 1000n$ is $o(n^2)$ because for any $c > 0$ we can select $n_0 = 1000/c + 1$ so that for all $n \geq n_0$, $cn^2 > cn(1000/c) = 1000n = f(n)$.

The strict lower bound little-ω is defined similarly.

Definition D.5

We say that $f(n)$ is $\omega(g(n))$ if for all $c > 0$ there exists a constant $n_0 > 0$ such that for all $n \geq n_0$ we have that $cg(n) < f(n)$.

Example D.6

The function $f(n) = n^2$ is $\omega(n)$ because for any $c > 0$ we can select $n_0 = c+1$ so that for all $n \geq n_0$ we have that $cn < n^2 = f(n)$.

D.6 COMMON ERRORS

There are a number of common errors that novices make with asymptotic notation. Some that you will want to avoid are listed below:

1. Asymptotic notation is about the growth rate of functions and not just the running time of an algorithm. Thus, for example, you should say that an algorithm has $O(n)$ *running time* and not simply that an algorithm is $O(n)$. Asymptotic notation can also be used to analyze memory, disk usage, number of database queries, and other properties of algorithms.

2. Never say "at least Big-O" since Big-O is an upper bound. Instead, use Ω or, better yet, Θ notation. When in doubt, be precise and use Θ notation unless you really don't have an exact characterization of the function.

3. While it is the case that $2n = O(n)$ and $3n = O(n)$, be wary of any circumstance in which you get something like $n + n + \ldots + n$ (n times) $= O(n)$ since this is not true.

D.7 CHAPTER SUMMARY AND KEY CONCEPTS

- **Asymptotic notation** allows us to compare the growth rate of functions. It is used extensively when analyzing the time and memory used by algorithms.

- **Theta (Θ) notation** is used to exactly capture the growth rate of a function. We say $f(n)$ is $\Theta(g(n))$ if there exists $n_0, c_1, c_2 > 0$ such that for all $n \geq n_0$ we have that $c_1 g(n) \leq f(n) \leq c_2 g(n)$.

- **Big-O (O) notation** is used to compute upper bounds on a function. We say $f(n)$ is $O(g(n))$ if there exists $n_0, c > 0$ such that for all $n \geq n_0$ we have that $f(n) \leq cg(n)$.

- **Big-Omega (Ω)** is used to compute lower bounds on a function. We say that $f(n)$ is $\Omega(g(n))$ if there exists $n_0, c > 0$ such that for all $n \geq n_0$ we have that $cg(n) \leq f(n)$.

- **Little-o (o) notation** is used to compute strict upper bounds on a function. We say that $f(n)$ is $o(g(n))$ if for all $c > 0$ there exists a constant $n_0 > 0$ such that for all $n \geq n_0$ we have that $f(n) < cg(n)$.

- **Little-omega (ω) notation** is used to compute strict lower bounds on a function. We say that $f(n)$ is $\omega(g(n))$ if for all $c > 0$ there exists a constant $n_0 > 0$ such that for all $n \geq n_0$ we have that $cg(n) < f(n)$.

EXERCISES

Prove the following by giving constants and then showing that the inequalities hold:

D.1 $n^2 = O(n^2)$

D.2 $3n^2 = O(n^3)$

D.3 $2n + 3 = O(n)$

D.4 $2n^2 + 4n = O(n^2)$

D.5 $3n^2 - 4n + 5 = O(n^2)$

D.6 $n^2 = \Omega(n^2)$

D.7 $7n^4 = \Omega(n^2)$

D.8 $3n + 4 = \Omega(n)$

D.9 $4n^2 - 3n = \Omega(n^2)$

D.10 $n^2 - 2n + 3 = \Omega(n^2)$

D.11 $n^3 = \Theta(n^3)$

D.12 $7n^2 = \Theta(n^2)$

D.13 $n + 8 = \Theta(n)$

D.14 $n^2 + 2n = \Theta(n^2)$

D.15 $4n^2 - 5n + 6 = \Theta(n^2)$

D.16 $n^2 = o(n^3)$

D.17 $100n^3 = o(n^4)$

D.18 $n^5 = \omega(n^4)$

D.19 $10n^3 = \omega(n^2)$

Prove the following properties.

D.20 If $f(n) = O(h(n))$ and $g(n) = O(h(n))$ then $f(n) + g(n) = O(h(n))$.

D.21 If $f(n) = O(g(n))$ and $g(n) = O(h(n))$ then $f(n) = O(h(n))$.

D.22 We have $f(n) = O(g(n))$ if and only if $g(n) = \Omega(f(n))$.

D.23 We have $f(n) = o(g(n))$ if and only if $g(n) = \omega(f(n))$.

D.24 Prove that $f(n) = \Theta(g(n))$ if and only if $f(n) = O(g(n))$ and $f(n) = \Omega(g(n))$.

D.25 Prove that any polynomial of degree d is $O(n^d)$. That is, for any $c_0, c_1, c_2, \ldots, c_d$ where $c_d > 0$, show that $\sum_{i=0}^{d} c_i n^i = O(n^d)$.

D.26 Prove that any polynomial of degree d is $\Theta(n^d)$. That is, for any $c_0, c_1, c_2, \ldots, c_d$ where $c_d > 0$, show that $\sum_{i=0}^{d} c_i n^i = \Theta(n^d)$.

D.27 Prove that for any base $b > 1$, $\log_b(n) = \Theta(\log_2(n))$. (Hint: Use the change of base formula $\log_b(a) = \frac{\log_d(a)}{\log_d(b)}$.)

D.28 Prove that for any $e > 0$, $\log_2(n) = O(n^e)$. (Hint: Choose $c = 1/e$ and use the fact that $\log_2(x) < x$ for all $x > 0$.)

Graphs

E.1 WHY YOU SHOULD CARE

Graphs are a data type that have many applications in computer science. They show up when we study computer networks and social networks and when your GPS finds the shortest route to your destination. In fact, there are so many applications of graphs in computer science that there are entire courses taught on the subject.

E.2 FORMAL DEFINITION

Simply put, a graph is collection of objects, called nodes or vertices, together with connections between some of them, called edges. While this sounds like a rather simple description, the study of graphs involves a large array of ideas and algorithms.

Example E.1

A group of students in a class can be modeled as a graph with each student being a node and edges between homework partners in the class.

An airline graph might consist of a set of cities (nodes) along with the pairs that have direct flights between them (edges).

Definition E.1

A graph is a pair $G = (V, E)$, where V is a set of vertices or nodes and E is the set of edges, or pairs of these vertices that are directly connected.

Example E.2

The following graph has three vertices and two edges:

It is represented by the tuple $G = (V, E)$, where the set of vertices is $V = \{A, B, C\}$ and the set of edges is $E = \{\{A, B\}, \{B, C\}\}$.

The above example is technically called an undirected graph. In some situations, we want the edges to be directed from one vertex to another. Some applications where this might be useful include one-way roads and relationships that are not symmetric (e.g., you could follow a celebrity on a social network, but they may not follow you back).

Definition E.2

A directed graph is a pair $G = (V, E)$, where V is a set of vertices or nodes and E is the set of edges, or ordered pairs of these vertices that are directly connected.

Example E.3

The following directed graph has three vertices and three edges:

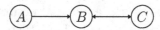

It is represented by the tuple $G = (V, E)$, where the set of vertices is $V = \{A, B, C\}$ and the set of edges is $E = \{(A, B), (B, C), (C, B)\}$. Note that the edges are now tuples (rather than sets) in which the order matters—for example, there is an edge from A to B but not from B to A. On the other hand, there are edges in both directions from B to C and from C to B.

E.3 GRAPH REPRESENTATION

There are two standard ways to represent graphs: adjacency lists and adjacency matrices.

The adjacency list representation of a graph maintains lists of vertices that each vertex is adjacent to, as in the following examples:

Example E.4

The graph in Example E.2 would be represented as follows:

```
A : [B]
B : [A, C]
C : [B]
```

That is, each vertex would be associated with a list of adjacent vertices.
The directed graph in Example D.3 would be represented by lists of all the vertices that each vertex has edges into:

```
A : [B]
B : [C]
C : [B]
```

The adjacency matrix representation of a graph uses a matrix or table to indicate when there is an edge between each vertex on the rows with each vertex on the columns, as seen in the following examples:

The graph in Example E.2 would be represented as follows:

	A	B	C
A	0	1	0
B	1	0	1
C	0	1	0

To tell if there is an edge between a pair of vertices, we look them up in the row and column of the above table and see if there is a 1 (an edge exists) or a 0 (no such edge exists). For example, the 1 in the second row and third column indicates that there is an edge between B and C. Note that the matrix is symmetric—if vertex u is adjacent to v, then v must also be adjacent to u. The directed graph in Example D.3 would record whether a vertex on the row has an edge into a vertex on the column:

	A	B	C
A	0	1	0
B	0	0	1
C	0	1	0

Note that the matrix is no longer symmetric as there is an edge from A to B (first row, second column), but not one from B to A (second row, first column).

The adjacency list and adjacency matrix representations each have their own advantages. The adjacency list might be a lot smaller in the case where most pairs of vertices don't have edges (e.g., social networks) as the total space usage is proportional to the number of vertices plus the number of edges ($O(|V|+|E|)$). An adjacency matrix on the other hand always uses the same amount of space equal to the square of the number of vertices ($O(|V|^2)$), which can get prohibitive when there are a large number of vertices.

On the other hand, adjacency matrices have very fast lookups. To see if a pair of vertices has an edge between them, we can just look up the matrix in constant ($O(1)$) time. The time to perform a similar operation on an adjacency list is proportional to the size of the list, which could be as large as the number of vertices in the graph ($O(|V|)$). Thus, which representation to pick depends on whether you have enough memory to store the matrix, failing which you have to deal with slower lookups.

E.4 GRAPH TERMINOLOGY

When working with graphs, it is important to know the terminology being used. All the following definitions are for undirected graphs, but similar analogs exist for directed graphs as well.

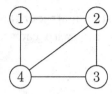

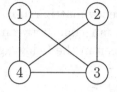

Figure E.1 An example graph   Figure E.2 A complete graph

Definition E.3

A vertex v is the **neighbor** of a vertex u if there is an edge between them (i.e., $\{u, v\} \in E$).
For example, the neighbors of vertex 1 in Figure E.1 are 2 and 4.

Definition E.4

The **degree** of a vertex is the number of its neighbors.
For example, the degrees of the vertices 1, 2, 3, and 4 in Figure E.1 are 2, 3, 2, and 3, respectively.

Definition E.5

A **complete graph** is one in which every pair of distinct vertices has an edge between them.
For example, the graph in Figure E.2 is a complete graph of size 4.

Definition E.6

A **path** is sequence of nodes with edges between consecutive nodes in the sequence.
For example, 2–4–1 is a path in the graph in Figure E.1 and 1–3–4–1–3 is a path in the graph in Figure E.2.

Definition E.7

A **simple path** is a path with no repeated nodes.
For example, 1–3–4–2 is a path in the graph in Figure E.2.

Definition E.8

A **cycle** is a path with the same first and last node.
For example, 1–3–4–1 is a cycle in the graph in Figure E.2.

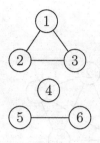

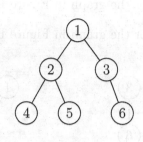

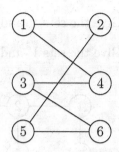

Figure E.3 A disconnected graph

Figure E.4 A tree

Figure E.5 A bipartite graph

Definition E.9

A graph is called **connected** if every pair of nodes has a path between them. For example, the graphs in Figures E.1, E.2, E.4, and E.5 are connected, but the one in Figure E.3 is not.

Definition E.10

A **tree** is a connected graph with no cycles.
For example, the graph in Figure E.4 is a tree but the one in Figure E.3 is not as it is disconnected and the one in Figure E.5 is not as it has a cycle (1–2–5–6–3–4–1).

Definition E.11

A **bipartite graph** is one whose vertices can be partitioned into two sets L and R (where $V = L \cup R$ and $L \cap R = \emptyset$) such that every edge in the graph is between a vertex in L and a vertex in R. For example, the graph in Figure E.5 is a bipartite graph with $L = \{1, 3, 5\}$ and $R = \{2, 4, 6\}$.

E.5 CHAPTER SUMMARY AND KEY CONCEPTS

- A **graph** is defined by a collection of **vertices** V with **edges** E between them.

- Graphs can be represented using **adjacency lists** or **adjacency matrices**.

- It is important to know graph terminology such as **neighbors, degree, complete, path, simple path, cycle, connected, tree,** and **bipartite**.

EXERCISES

E.1 Give the sets V and E for the graph in Figure E.3.

E.2 Give the sets V and E for the graph in Figure E.4.

E.3 Give the sets V and E for the graph in Figure E.5.

E.4 Give the sets V and E for the graph in Figure E.6.

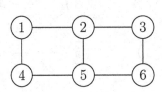

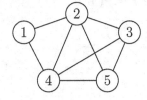

Figure E.6 Exercise graph 1   **Figure E.7** Exercise graph 2

E.5 Give the sets V and E for the graph in Figure E.7.

E.6 Give the adjacency list representation for the graph in Figure E.3.

E.7 Give the adjacency list representation for the graph in Figure E.4.

E.8 Give the adjacency list representation for the graph in Figure E.5.

E.9 Give the adjacency list representation for the graph in Figure E.6.

E.10 Give the adjacency list representation for the graph in Figure E.7.

E.11 Give the adjacency matrix representation for the graph in Figure E.3.

E.12 Give the adjacency matrix representation for the graph in Figure E.4.

E.13 Give the adjacency matrix representation for the graph in Figure E.5.

E.14 Give the adjacency matrix representation for the graph in Figure E.6.

E.15 Give the adjacency matrix representation for the graph in Figure E.7.

E.16 List the degree of each node for the graph in Figure E.6.

E.17 List the degree of each node for the graph in Figure E.7.

E.18 A subgraph of a graph is a subset of the vertices and all the edges between them. What is the size of the largest subgraph for the graph in Figure E.6 that is a complete graph? What is the size of the largest subgraph for the graph in Figure E.7 that is a complete graph?

E.19 Find a longest cycle with only the first and last vertices repeated in the graph in Figure E.6.

E.20 Find a longest cycle with only the first and last vertices repeated in the graph in Figure E.7.

E.21 Is the graph in Figure E.6 a tree?

E.22 Is the graph in Figure E.7 a tree?

E.23 Is the graph in Figure E.6 bipartite? List out the vertices in L and R if it is.

E.24 Is the graph in Figure E.7 bipartite? List out the vertices in L and R if it is.

Loop Invariants

F.1 WHY YOU SHOULD CARE

When you write a function to search or sort a list, how do you know that it will work on all valid inputs? You certainly can't test it on every possible input. In computer science, we prove correctness of our algorithms. A common technique for proving correctness of algorithms with loops is the loop invariant proof. Read on to learn more!

F.2 SUMMING A LIST

Consider the following algorithm (Algorithm F.1) for summing up a list of numbers:

Algorithm F.1: sum(L)

> **input** : A list of numbers L of size $|L|$
> **output:** The sum of the elements of L
> 1: $total = 0$
> 2: **for** $i = 0, \ldots, |L| - 1$ **do**
> 3: $total = total + L[i]$
> 4: **return** $total$

This is a very simple algorithm, but it makes a great first example to see how we can prove that it always works. To prove that the sum algorithm is correct, we introduce a loop invariant.

Definition F.1

A loop invariant is a statement that is true prior to the start of a loop and remains true at the start of each iteration of the loop. Combined with the termination condition of the loop, it tells us that the algorithm did in fact solve the problem correctly.

A loop invariant for the sum problem is given below. Note that for the rest of this chapter we will use the notation $L[i : j]$ to indicate the sub-list consisting of

DOI: 10.1201/9781003383284-F

$L[i], L[i+1], \ldots, L[j-1]$ (i.e., starting from the ith element up to but not including the jth element), similar to the notation used by Python for list slices.

Example F.1

Loop invariant: At the start of iteration i, *total* is equal to the sum of the list $L[0:i]$.

We use loop invariants such as this one to essentially write a proof by induction to show correctness of the algorithm. We start with the loop initialization (similar to the base step) and then show that the invariant is maintained (similar to the inductive step) after each iteration. Unlike a proof by induction that shows the result for all positive n, we terminate our proof based on the termination condition of the loop.

Example F.2

At the **initialization** step, we consider what happens when $i = 0$. At this stage we don't have any elements in our list since we don't include the $i = 0$ element. The sum of an empty list is always zero and we correctly have *total* equal to zero at initialization on Line 1.

Example F.3

At the **maintenance** step, we have to show that each iteration makes progress toward our goal. Similar to induction, we will do this by proving that if the loop invariant holds at the start of the jth iteration, it will also hold after the $(j+1)$st iteration.

We assume that the loop invariant holds for some iteration $j \geq 0$. That is, at the start of iteration j, *total* holds the sum of $L[0:j]$. During iteration j, we only add $L[j]$ into *total* on Line 3. Thus, at the start of iteration $(j+1)$, we have that *total* is equal to the sum of $L[0:j+1]$, as desired.

Example F.4

In the **termination** step, we have to figure out when the loop will terminate and then apply that termination value to the loop invariant to show that the loop accomplished what we set out to do.

The loop for the sum algorithm will sum indexes up through $|L| - 1$ and will thus terminate when $i = |L|$. Substituting this into the loop invariant, we get that at the termination of the loop of the sum function, we have that *total* is equal to the sum of the list $L[0:|L|]$, which is precisely the entire list. Thus, at the end of the loop, *total* holds the sum of the entire list L.

Every loop invariant proof needs all of the above four parts: statement of the loop invariant, initialization, maintenance, and termination. We'll see some more examples next.

F.3 EXPONENTIATION

Consider the standard algorithm for exponentiation (Algorithm F.2) below. It very straightforwardly just uses an accumulator to compute the exponent. Think about what would be the loop invariant in this case.

Algorithm F.2: $\exp(a, n)$

input : A real number a and a positive integer n
output: The value a^n
1: $result = 1$
2: **for** $i = 1, \ldots, n$ **do**
3:     $result = result * a$
4: **return** $result$

The invariant involves the *result* variable that starts at 1 and should end at a^n. At each iteration, the value of *result* is multiplied by a. Guided by this, we propose the following invariant:

Example F.5

Loop invariant: At the start of iteration i, *result* is equal to a^{i-1}.

The rest of the proof looks as follows.

Example F.6

At the **initialization** step, before the start of the loop, the value of *result* is initialized to be 1 on Line 1. We can verify that before the start of the first iteration (when $i = 1$), *result* is indeed equal to $a^{i-1} = a^{1-1} = a^0 = 1$.

Example F.7

For the **maintenance** step, we assume that the invariant holds for some iteration $j \geq 1$. That is, at the start of iteration j we have that $result = a^{j-1}$. During iteration j, we multiply *result* by a on Line 3, so its new value will be $a^j = a^{(j+1)-1}$ at the start of iteration $(j + 1)$, as desired.

Example F.8

In the **termination** step, we see that the loop terminates when $i = n + 1$. Substituting this into the invariant, we get that at the termination of the loop the value of *result* is $a^{(n+1)-1} = a^n$, as desired. Thus, the function does correctly exponentiate a^n.

The correctness of the previous algorithm might have looked obvious to you, so let us see how we can write a proof for a more interesting exponentiation algorithm

(Algorithm F.3). It will also show us an example that has a while loop rather than a for loop.

Algorithm F.3: fastexp(a, n)

   input : A real number a and a positive integer n
   output: The value a^n
  1: $result = 1$
  2: $b = a$
  3: $e = n$
  4: **while** $e > 0$ **do**
  5:    **if** e is odd **then**
  6:       $result = result * b$
  7:       $e = e - 1$
  8:    **else**
  9:       $b = b * b$
10:       $e = e/2$
11: **return** $result$

This algorithm works similarly to the previous one, accumulating the product into an accumulator variable, except that if the exponent is even it performs the following speedup: it squares the base and halves the exponent. This considerably decreases the number of multiplications needed since we often halve the exponent, earning the algorithm the name **fast exponentiation**.

Example F.9

The following is a trace of the variables at the end of each iteration when computing the value fastexp$(2, 31)$:

iteration	0	1	2	3	4	5	6	7	8	9
$result$	1	2	2	2^3	2^3	2^7	2^7	2^{15}	2^{15}	2^{31}
b	2	2	2^2	2^2	2^4	2^4	2^8	2^8	2^{16}	2^{16}
e	31	30	15	14	7	6	3	2	1	0

At the start of the while loop (iteration 0), the values of $result$, b, and e are the initialized values 1, 2, 31, respectively. Whenever we encounter an odd exponent e (at the end of iterations 0, 2, 4, 6, and 8), we simply multiply the base b into $result$ and decrement the exponent by one. However, when we encounter an even exponent (at the end of iterations 1, 3, 5, and 7) we shorten the rest of the computation by halving the exponent and squaring the base since $b^e = (b^2)^{e/2}$.

To prove correctness of the algorithm, we need to find a pattern in the values of $result$, b, and e. Notice that at every iteration the remaining product that needs to be multiplied into the $result$ variable is b^e. After some thought, you might get the following invariant:

Example F.10

Loop invariant: At the start of any iteration, $result \cdot b^e = a^n$.

For example, in the table in the previous example, you should verify that $result \cdot b^e$ is always equal to 2^{31}. Notice that this invariant doesn't have an iteration number as we had in the previous `for` loop examples.

Example F.11

At the **initialization** step, before the start of the loop (Lines 1-3), the value for $result$, b, and e are initialized to 1, a, and n respectively. It is easy to see that $result \cdot b^e = 1 \cdot a^n = a^n$.

Example F.12

For the **maintenance** step, we assume that the invariant holds true for a given iteration and show that it is still true in the next iteration. That is, we assume that at the start of an iteration we have $result \cdot b^e = a^n$. We now have two cases: when e is odd and when e is even.

If e is odd (Lines 6-7), we multiply $result$ by b and decrement e by 1. This causes $result$ to go up by a factor of b and b^e to go down by a factor of b. These cancel each other out and we end up with $result \cdot b^e$ being unchanged (still a^n) as desired.

If e is even (Lines 9-10), we square b and halve e. Since $b^e = (b^2)^{e/2}$, this means that b^e and thus $result \cdot b^e$ remains unchanged (still a^n) as needed.

Thus, the invariant holds true at the start of the next iteration.

Example F.13

In the **termination** step, we see that the `while` loop terminates when $e = 0$ (Line 4). Substituting this into the invariant, we get that $result \cdot b^0 = a^n$ or that $result = a^n$, as we wanted to show.

F.4 INSERTION SORT

The following Insertion Sort algorithm (Algorithm F.4) sorts a list of comparable types (i.e., types that have the $>$ operation defined). It works much in the same way that most people sort playing cards in their hands: as we encounter a card (starting from the left), we insert that card in the correct position for the cards encountered thus far. More precisely, during each iteration i, the algorithm inserts the ith element in the correct position for the sublist $L[0 : i + 1]$. At each successive iteration a larger sublist of the list is sorted. Eventually, when all the items in the list have been encountered, the entire list will be sorted.

Algorithm F.4: InsertionSort(L)

input : A list L of size $|L|$ with comparable elements
output: The list L sorted
1: **for** $i = 1, \ldots, |L| - 1$ **do**
2:     $value = L[i]$
3:     $j = i - 1$
4:     **while** $j \geq 0$ and $L[j] > value$ **do**
5:         $L[j + 1] = L[j]$
6:         $j = j - 1$
7:     $L[j + 1] = value$
8: **return** L

Example F.14

Consider how Insertion Sort works on a list $[3, 1, 4, 1, 5]$ after each iteration i:

- ($i = 1$) $[\mathbf{1}, \mathbf{3}, 4, 1, 5]$ The 1 at $L[1]$ gets inserted before the 3.

- ($i = 2$) $[\mathbf{1}, \mathbf{3}, \mathbf{4}, 1, 5]$ The 4 at $L[2]$ gets inserted at its current position.

- ($i = 3$) $[\mathbf{1}, \mathbf{1}, \mathbf{3}, \mathbf{4}, 5]$ The 1 at $L[3]$ gets inserted at $L[1]$.

- ($i = 4$) $[\mathbf{1}, \mathbf{1}, \mathbf{3}, \mathbf{4}, \mathbf{5}]$ The 5 at $L[4]$ gets inserted at its current position.

Note that as we go further into the list, we get a sorted prefix list (i.e., the elements in bold) until the entire list is sorted.

We will formally prove the correctness of Insertion Sort next. Notice that each iteration of Insertion Sort has a larger prefix of the list sorted (indicated in bold in the above example). We will use this idea to develop the following loop invariant for it:

Example F.15

A **loop invariant** for insertion sort is:

> At the start of iteration i, the sub-list $L[0 : i]$ consists of the original elements of $L[0 : i]$ in sorted order.

Example F.16

For **initialization**, we have that when $i = 1$ the list $L[0 : 1]$ consists of a single element which is the original element $L[0]$ and a list of size one is always sorted.

Example F.17

For **maintenance**, we assume that the loop invariant is true for some $k \geq 1$. That is, at the start of iteration k, the sub-list $L[0 : k]$ consists of the original elements of $L[0 : k]$ in sorted order. We will show that the loop invariant is true at the start of iteration $k + 1$. During iteration k, the algorithm copies the kth item into *value* and the `while` loop moves all the items before position k that are larger than *value* up by one position until it finds the correct position for *value* and the algorithm then inserts *value* into this position (Lines 4-7). Since this operation only moves the position of $L[k]$ within the sub-list $L[0 : k + 1]$, the list $L[0 : k + 1]$ consists of the original elements of $L[0 : k + 1]$. Moreover, since $L[0 : k]$ was previously in sorted order and the above insertion operation places $L[k]$ in the correct position, the sub-list $L[0 : k + 1]$ is now in sorted order. Thus, the loop invariant is maintained.

Example F.18

For **termination**, the `for` loop terminates when $i = |L|$. Substituting this into the loop invariant, we get that at termination the sub-list $L[0 : |L|]$, which is the entire list L, consists of the original elements of L in sorted order. Thus, at termination the list L is correctly sorted.

Note that the above proof could be made even more rigorous by adding in a separate loop invariant proof for the `while` loop. This is left as an excercise for the reader.

F.5 CHAPTER SUMMARY AND KEY CONCEPTS

- **Loop invariants** are statements that are true at the start of each iteration of a loop. They allow us to prove correctness of the algorithm.

- Any loop invariant proof has four parts: (1) **stating** the invariant, (2) an **initialization** step in which the invariant is shown to be true at the start of the loop, (3) a **maintenance** step in which it is shown that if the invariant was true before the previous iteration then it must be true before the next iteration, and (4) a **termination** step in which the invariant is used to show that the loop correctly computed what it was suppsed to.

EXERCISES

Write out algorithms (with **loops**) for each of the following and prove them correct using loop invariant proofs.

F.1 Finding the maximum of a list

F.2 Finding the minimum of a list

F.3 Summing up the first n positive integers

F.4 Summing up the first n powers of 2

F.5 Computing the factorial of n $(1 \times 2 \times 3 \ldots \times n)$

F.6 Computing the nth Fibonacci number (iterative)

F.7 Linear search of a list

F.8 Reversing a list by creating a new list

F.9 Reversing a list in place

F.10 Selection sort (Hint: First prove a loop invariant for the inner loop, then use this to prove a loop invariant for the outer loop.)

F.11 Bubble sort (Hint: First prove a loop invariant for the inner loop, then use this to prove a loop invariant for the outer loop.)

F.12 Binary search (iterative)

Recurrence Relations

G.1 WHY YOU SHOULD CARE

Recurrence relations, sometimes called recurrences, are a way of analyzing recursive algorithms to compute their run-time. Since recursion, particularly divide-and-conquer, comes up quite often in algorithm design, knowing how to solve them is important for computer scientists.

G.2 MERGE SORT

Divide and conquer is a recursive algorithmic technique in which the instance of the problem being solved is broken down into two or more smaller versions of the original problem and the answer to these helps compute the answer to the original problem. An important example of this is the Merge Sort algorithm.

The Merge Sort algorithm sorts a list by splitting the list into two close-to-equal halves, sorting both halves recursively, and then merging the halves together. (Since the length of the list could be odd, one of the halves might be one larger than the other, hence the "close-to-equal.") In the base case, when the list has length one, we don't have to do anything as a list of size one is already sorted. We won't delve into the details of the algorithm here and will instead just analyze its run-time.

Let the run-time of Merge Sort on a list of size n be denoted by $T(n)$ for all $n \geq 1$. From the above description, we can see that the running time of the entire algorithm will be the time to sort two half-sized lists and the time to merge these lists. The time to sort each half-sized list can be denoted by $T(n/2)$ based on the definition of T. For simplicity, we'll assume that n is always a power of two to avoid complications of unequal halves. The time to merge the lists together turns out to be linear in n and, for simplicity, we just denote it as cn for some constant $c > 0$. Lastly, the base case of Merge Sort is when the list has size one, in which case we simply return the list with zero additional work done because a list of size one is already sorted. Putting all this together, we get the following definition of T, called a **recurrence relation** or **recurrence**:

$$T(n) = \begin{cases} 0 & \text{if } n = 1 \\ 2T(n/2) + cn & \text{if } n > 1. \end{cases}$$

DOI: 10.1201/9781003383284-G

We'll see next a few different ways to solve such a recurrence relation.

G.3 RECURSION TREE METHOD

One way to solve the above recurrence relation, called the recursion tree method, is to use the formula for T and expand it a few times like this:

$$
\begin{aligned}
T(m) = & & 2T(m/2) + cm & & \text{(using the formula for } T \text{ when } n = m) \\
= & & 2\left(2T(m/4) + cm/2\right) + cm & & \text{(using the formula for } T \text{ when } n = m/2) \\
= & & 4T(m/4) + 2cm & & \text{(simplifying with algebra)} \\
= & & 4\left(2T(m/8) + cm/4\right) + 2cm & & \text{(using the formula for } T \text{ when } n = m/4) \\
= & & 8T(m/8) + 3cm & & \text{(simplifying with algebra).}
\end{aligned}
$$

We can continue this process a few more times to see that this seems to be following the pattern (after expanding the formula k times):

$$
T(m) = 2^k T(m/2^k) + kcm.
$$

We can continue the above expansion until such time as we have reduced the problem down to size one. This happens precisely when $m/2^k = 1$ or when $m = 2^k$ or $k = \log_2(m)$. Substituting this value of k into the above formula, we get

$$
\begin{aligned}
T(m) = & & 2^{\log_2(m)} T(m/2^{\log_2(m)}) + cm\log_2(m) \\
= & & mT(m/m) + cm\log_2(m) & & (\text{as } 2^{\log_2(m)} = m) \\
= & & mT(1) + cm\log_2(m) \\
= & & cm\log_2(m) & & (\text{as } T(1) = 0).
\end{aligned}
$$

We can conclude from this derivation that the run-time of Merge Sort is then $cm\log_2(m)$, more simply expressed (using Big-O notation, see Appendix D) as $O(n\log(n))$.

G.4 A REVIEW OF SOME LOG RULES

As you just saw in the previous section, you will need to be comfortable working with logarithms in this chapter. Here are a few rules that are worth remembering. For any $b > 1$ and $n, m > 0$, we have

$$
\begin{aligned}
\log_b(1) &= 0 \\
\log_b(b) &= 1 \\
b^{\log_b(n)} &= n \\
\log_b(nm) &= \log_b(n) + \log_b(m) \\
\log_b(n/m) &= \log_b(n) - \log_b(m) \\
\log_b(n^m) &= m\log_b(n).
\end{aligned}
$$

G.5   SUBSTITUTION METHOD

Another, simpler, way to prove the recurrence can be to use strong induction. For this method, we need to know what the formula is beforehand. We can write the following induction hypothesis and prove it inductively.

For any $n \geq 1$,

$$P(n) : T(n) = cn \log_2(n).$$

For the base case, we have that when $n = 1$, we know that $T(1) = 0$ and $cn \log_2(n) = 0$ since $\log_2(1) = 0$. Thus $P(1)$ holds.

We fix some $k > 1$ and assume that the induction hypothesis holds for all values $n < k$. We will show that the induction hypothesis holds for $n = k$ as well.

When $n = k$, we have

$$
\begin{aligned}
T(k) &= & 2T(k/2) + ck & \quad \text{(by the definition of } T) \\
&= & 2(c(k/2) \log_2(k/2)) + ck & \quad \text{(by the induction hypothesis when } n = k/2) \\
&= & ck \log_2(k/2) + ck & \quad \text{(simplifying with algebra)} \\
&= & ck \log_2(k) - ck + ck & \quad \text{(as } \log_2(k/2) = \log_2(k) - \log_2(2) = \log_2(k) - 1 \text{)} \\
&= & ck \log_2(k), &
\end{aligned}
$$

thus proving the inductive step. We can wrap up the strong induction proof as usual.

G.6   ANALYZING THE KARATSUBA-OFMAN ALGORITHM

There is another divide and conquer algorithm for multiplying large numbers called the Karatsuba-Ofman algorithm with the following recurrence:

$$
T(n) = \begin{cases} 0 & \text{if } n = 1 \\ 3T(n/2) + cn & \text{if } n > 1. \end{cases}
$$

We'll see how to analyze it using the above two techniques next. Using the recursion tree method, we get:

$$
\begin{aligned}
T(m) &= & 3T(m/2) + cm & \quad \text{(formula for } T \text{ when } n = m) \\
&= & 3\left(3T(m/4) + cm/2\right) + cm & \quad \text{(formula for } T \text{ when } n = m/2) \\
&= & 9T(m/4) + (3/2)cm + cm & \quad \text{(simplifying with algebra)} \\
&= & 9\left(3T(m/8) + cm/4\right) + (3/2)cm + cm & \quad \text{(formula for } T \text{ when } n = m/4) \\
&= & 27T(m/8) + (3/2)^2 cm + (3/2)cm + cm & \quad \text{(simplifying with algebra).}
\end{aligned}
$$

We can continue this procedure a few more times to see that this seems to be following the pattern (after expanding the formula k times):

$$
T(m) = 3^k T(m/2^k) + \sum_{i=0}^{k-1} (3/2)^i cm.
$$

We can continue this until we have reduced the problem down to size one. This happens precisely when $m/2^k = 1$ or when $m = 2^k$ or $k = \log_2(m)$. Substituting this value of k into the above formula, we get

$$T(m) = 3^{\log_2(m)} T(m/2^{\log_2(m)}) + \sum_{i=0}^{\log_2(m)-1} (3/2)^i cm$$

$$= 3^{\log_2(m)} T(m/m) + \sum_{i=0}^{\log_2(m)-1} (3/2)^i cm \qquad \text{(as } 2^{\log_2(m)} = m\text{)}$$

$$= 3^{\log_2(m)} T(1) + \sum_{i=0}^{\log_2(m)-1} (3/2)^i cm$$

$$= \sum_{i=0}^{\log_2(m)-1} (3/2)^i cm \qquad \text{(as } T(1) = 0\text{)}$$

$$= \frac{(3/2)^{\log_2(m)} - 1}{3/2 - 1} cm \qquad \text{(formula for geometric sum)}$$

$$= 2cm(3^{\log_2(m)}/2^{\log_2(m)} - 1) \qquad \text{(dividing by } 1/2 \text{ is the same as}$$
$$\text{multiplying by 2)}$$

$$= 2cm(3^{\log_2(m)}/m - 1) \qquad \text{(as } 2^{\log_2(m)} = m\text{)}$$

$$= 2cm^{\log_2 3} - 2cm \qquad \text{(logarithm rules).}$$

We can conclude from this derivation that the run-time of the algorithm is thus $O(n^{\log_2 3})$ or $O(n^{1.585})$.

The substitution method for this proof proceeds as follows. For any $n \geq 1$,

$$P(n) : T(n) = 2cn^{\log_2 3} - 2cn.$$

For the base case, we have that when $n = 1$, we know that $T(1) = 0$ and $2cn^{\log_2 3} - 2cn = 2c - 2c = 0$. Thus $P(1)$ holds.

We fix some $k > 1$ and assume that the induction hypothesis holds for all values $n < k$. We will show that the induction hypothesis holds for $n = k$ as well.

When $n = k$, we have

$$T(k) = \qquad 3T(k/2) + ck \qquad \text{(by the definition of } T\text{)}$$

$$= \quad 3(2c(k/2)^{\log_2 3} - 2ck/2) + ck \quad \text{(by the induction hypothesis when } n = k/2\text{)}$$

$$= \qquad 6ck^{\log_2 3}/2^{\log_2 3} - 3ck + ck \qquad \text{(algebra)}$$

$$= \qquad 2ck^{\log_2(3)} - 2ck \qquad \text{(as } 2^{\log_2(3)} = 3\text{)},$$

thus showing the inductive step. We can again wrap up the strong induction proof as usual.

G.7   CHAPTER SUMMARY AND KEY CONCEPTS

- **Recurrences** are recursive formulas, usually used to express the running time of a recursive algorithm.

- The **recursion tree method** can be used to unroll a recurrence to get an exact formula for it.

- The **substitution method** uses strong induction to formally prove the formula for the recurrence.

EXERCISES

Solve the following recurrences using the recursion tree method. In each case, assume that $T(1) = 0$.

G.1 $T(n) = 2T(n/4) + n$

G.2 $T(n) = 2T(n/2) + n^2$

G.3 $T(n) = 7T(n/8) + n$

G.4 $T(n) = 3T(n/4) + n^2$

G.5 $T(n) = 5T(n/8) + n^3$

Solve the following recurrences using the substitution method (strong induction). In each case, assume that $T(1) = 0$. (Hint: Use the recursion tree method first to get an induction hypothesis in each case.)

G.6 $T(n) = 2T(n/4) + n$

G.7 $T(n) = 2T(n/2) + n^2$

G.8 $T(n) = 7T(n/8) + n$

G.9 $T(n) = 3T(n/4) + n^2$

G.10 $T(n) = 5T(n/8) + n^3$

Further Reading

There are a number of books on discrete math for computer science available, see for example [Epp19, FK03, GKP89, LN22, LZ19, SDB10]. However, I believe this is the first book that uses theory of computation as the way to introduce these discrete math topics.

Since this book is intended to be an introduction to the mathematical ideas in computer science first and only a light introduction to the theory of computation, there are many topics in the latter that have been very briefly discussed and others that have been omitted entirely. For those wishing to read more about these topics, I recommend some of the standard texts by Hopcroft, Motwani, Ulman [HMU06], Martin [Mar90], and Sipser [Sip96]. For a more advanced look at the topic one could look at the Complexity Theory Companion [HO02]—some of my treatment of topics was heavily influenced by what I have learned from the authors of this book.

For students looking to brush up on some of the basic computer science prerequisites in this book, I recommend the Discovering Computer Science book by Havill [Hav20].

I often like to delve into the philosophy of computer science with my students. For this, I like to use the Stanford Encyclopedia of Philosophy [sep24].

For more detail on the algorithmic topics, especially loop invariants and recurrences, I recommend the algorithms bible by Cormen, Leiserson, Rivest, and Stein [CLRS09].

To learn more about probability in computer science, please see [MU05]. The book by Motwani and Raghavan [MR95] is the standard text on randomized algorithms.

Bibliography

[CLRS09] Thomas H. Cormen, Charles E. Leiserson, Ronald L. Rivest, and Clifford Stein. *Introduction to Algorithms, Third Edition*. The MIT Press, 3rd edition, 2009.

[Epp19] Susanna Samuels Epp. *Discrete Mathematics with Applications*. Cengage Learning, 5th edition, 2019.

[FK03] T. Feil and J. Krone. *Essential Discrete Mathematics for Computer Science*. Prentice Hall, 2003.

[GKP89] Ronald L. Graham, Donald E. Knuth, and Oren Patashnik. *Concrete Mathematics: A Foundation for Computer Science*. Addison-Wesley, 1989.

[Hav20] Jessen Havill. *Discovering Computer Science: Interdisciplinary Problems, Principles, and Python Programming (Second Edition)*. Chapman and Hall/CRC Press, 2nd edition, 2020.

[HMU06] J. E. Hopcroft, R. Motwani, and J. D. Ullman. *Introduction to Automata Theory, Languages, and Computation*. Addison Wesley, 3rd edition, 2006.

[HO02] Lane A. Hemaspaandra and Mitsunori Ogihara. *The Complexity Theory Companion*. Springer-Verlag, 2002.

[LN22] David Liben-Nowell. *Connecting Discrete Mathematics and Computer Science*. Cambridge University Press, 2nd edition, 2022.

[LZ19] Harry Lewis and Rachel Zax. *Essential Discrete Mathematics for Computer Science*. Princeton University Press, 2019.

[Mar90] John C. Martin. *Introduction to Languages and the Theory of Computation*. McGraw-Hill, Inc., USA, 1st edition, 1990.

[MR95] Rajeev Motwani and Prabhakar Raghavan. *Randomized Algorithms*. Cambridge University Press, 1995.

[MU05] Michael Mitzenmacher and Eli Upfal. *Probability and Computing: Randomized Algorithms and Probabilistic Analysis*. Cambridge University Press, 2005.

[SDB10]   Cliff Stein, Robert Drysdale, and Kenneth Bogart. *Discrete Mathematics for Computer Scientists*. Addison-Wesley Publishing Company, 2010.

[sep24]   Stanford Encyclopedia of Philosophy. `https://plato.stanford.edu`, 2024.

[Sip96]   Michael Sipser. *Introduction to the Theory of Computation*. International Thomson Publishing, 1st edition, 1996.

Index

λ-transition, 56

adjacency list, 188
adjacency matrix, 189
Alan Turing, 163
alphabet, 13
asymptotic notation, 181

base step, 88
biconditional, 40
Big-O notation, 41, 182
Big-Omega (Ω) notation, 183
binary, 30, 166
Binary Search algorithm, 102, 201
bipartite graph, 191
Bubble Sort algorithm, 201

Cantor, 119
cardinality, 3
Cartesian product, 10
Church-Turing thesis, 152
class (of languages), 161
closure properties, 82, 134
codomain, 12
codon, 27
complement, 6
complete graph, 190
computable, 154
computational biology, 27
concatenation, 13
conditional, 38
congruence, 177
connected, 190
context-free grammar, 130, 133
context-free language, 134
contradiction, 117
contrapositive, 39
converse, 39
cycle, 190

De Morgan's Laws, 6, 37
decidable, 154
decimal, 30
definition, 77
degree, 190
deterministic finite automaton, 23
DFA, 23
Diagonal Problem D, 155
diagonalization, 119, 156
direct proof, 77
divisible, 30, 177
division algorithm, 177
divisor, 177
domain, 12

edge, 187
element (of a set), 2
empty set, 2
empty string, 13
Euclid's algorithm, 179
even, 77
event, 171
    complement, 171
    independent, 172
    mutually exclusive, 172
exponentiation algorithm, 196

factor, 177
factorial, 87, 167, 201
fast exponentiation, 197
Fibonacci numbers, 97, 201
floor division, 176
function, 12
    codomain, 12
    domain, 12
fundamental theorem of arithmetic, 118

grammar, 130, 133
graph, 187